The Ground Beneath Our Work

Nature-Informed Therapy and Care for a World in Need

Heidi Schreiber-Pan, Ph.D., LCPC
Founder and Executive Director,
Center for Nature Informed Therapy

Chesapeake Publication
Baltimore, Maryland

First edition, 2026

Hardcover ISBN 978-1-962949-17-0
Paperback ISBN 978-1-962949-16-3

Published by Chesapeake Publication
Baltimore, Maryland

www.natureinformedtherapy.org

Interior design and illustrations: Xiao Pan

ROOTED™ is a trademark of the Center for Nature Informed Therapy.

Printed in the United States of America

Praise for

The Ground Beneath Our Work

"**This book fills a critical need in clinical practice:** a framework that is both grounded in nature and rooted in justice. Dr. Schreiber-Pan's integrative approach — from nature as co-therapist to the ROOTED model and the 30-day clinician toolkit — gives clinicians tangible pathways to enhance regulation, expand sensory engagement, and cultivate reciprocity with their clients. A deeply human and clinically usable resource."
— **Dr. Christine Lynn Norton**

Professor of Social Work, Texas State University

"**Rooted in lived clinical wisdom and a deep reverence for the natural world,** this book offers a guide that is both lyrical and practical—an inviting blend of reflection and grounded tools for integrating nature into the counseling process. Dr. Heidi Schreiber-Pan writes with clarity and warmth, offering insights, practices, and rich vignettes that will resonate with both new and seasoned helping professionals. A vital companion for anyone seeking to bring the natural world into the healing journey."
— **Ryan F. Reese, PhD**

Author of *Natural Approaches to Optimal Wellness: Integrating EcoWellness into Counseling*

"**Grounded in experience, research, and psychological insight,** The Ground Beneath Our Work offers a vision of healing through relationship with the natural world. This book is both a practical guide and a gentle call to remember our deep connection with nature, offering inspiration for anyone seeking to weave nature more fully into life and practice."
— **Megan Delaney, PhD**

Author of *Nature is Nurture: Counseling and the Natural World*

"In The Ground Beneath Our Work Schreiber-Pan illuminates a way of being human that is truly meant for every human being... We can re-ground ourselves and find nourishment and healing not just in far flung wild places, but on our neighborhood sidewalk or our own back porch... this book has forever changed how I sit in my own backyard... to find continued connection to others and my place in the world."

— Thomas R. Medema

Former National Park Service Associate Director for Interpretation, Education, and Volunteers

"The Ground Beneath Our Work... is a must-read for clinicians and thought leaders who care deeply about the environment... It offers both depth and practicality, and it is written with remarkable beauty, clarity, and precision... [It] will spark motivation, reasoning, and provide a clear path forward in your practice."

— Jennifer Udler, LCSW-C, SEP

Author, Walk and Talk Therapy: *A Clinician's Guide to Incorporating Movement and Nature into Your Practice*

Foreword

The Ground Beneath Our Work is a rare book: it speaks with equal warmth and usefulness to licensed clinicians and to the wider circle of helpers who care for others in schools, hospitals, parks, congregations, communities, and homes. While reading, I found myself having two repeated reactions: "I didn't know that," and "Of course—that makes sense." I also caught myself wishing I had encountered these insights earlier, especially before I wrote my most recent work.

One of the strengths of Dr. Heidi Schreiber-Pan's writing is that she does not argue people into caring about the natural world; she invites them. Through story, clarity, and practical guidance, she helps readers—especially those unfamiliar with or hesitant about Nature-Informed work—discover that the path is not complicated, and that it has been available to us all along.

As I moved through these pages, I kept remembering moments from my own clinical life when "the room" itself made the difference. I think of a well-known author and spiritual guide who came to see me during a season of persistent anxiety. Our work was steady, but it felt constrained—too contained, too head-heavy. On an impulse, I suggested that we leave my office and take our session to a nearby lake.

As we walked, something softened. The fresh air, the music of birds, the wind moving through trees, and the presence of water created space—space that neither of us could manufacture inside four walls. Mid-stride, he stopped and asked, with real fear, "Will these panic attacks ever end?" And I felt an unusual freedom to respond with equal honesty: "Oh, without a doubt. That isn't your main problem." Startled, he asked, "Then what is?" I said, "The most pressing question is whether you will learn what you need to learn—how your life might deepen and change— before the symptoms fade and you return to the old status quo."

He paused, looked out over the water, and breathed. A bird lifted from the ground into a high branch, and he watched it as if it were an answer. It was, in its own way, *just right.*

Most people can imagine the ocean easing stress, waves doing what waves do. But Dr. Schreiber-Pan invites us to experience much more than relaxation. She offers a way of practicing and living that treats the natural world not as background scenery, but as an active partner in care. The book is rich with what I would call the "lyrics" of good work: concrete techniques, ethical considerations, and practical ways to integrate Nature-Informed support across settings. Yet it also offers the "music": an outlook—an orientation—that makes you want to do the work more consciously, more relationally, and with greater respect for what nature contributes.

And this matters, because the pace and pressure of helping work can be relentless. I remember a period when many things had been going well—until, suddenly, they were not. Family medical concerns emerged. People I cared about were in crisis. And, as often happens, others sought me out for guidance precisely because I was "the helper." In the middle of that convergence of stress, I looked out a window and saw two views at once: an area of land made brown by "management," and beyond it, a lake with geese on the water and two blue herons gliding above it. Those two images, harm and beauty, erosion and grace, brought me up short.

I did what my own nervous system already knew to do: I stopped, went outside, and walked through the landscapes of my life. When I returned, I could hold both realities at once, storms and blue skies—and I sat quietly, breathing, letting the day settle. In that moment, I realized something that Dr. Schreiber-Pan makes beautifully clear throughout this book: Nature-Informed work is not for some of us, only some of the time. It is a way of remembering what is always true—that we are shaped by relationship, and that the more-than-human world can help us live and serve with greater clarity, steadiness, and humility.

As you read what follows, I hope you will feel invited to step out of the closed room—out of the harsh lighting and the shrinking of attention—and into a wider field of care. *The Ground Beneath Our Work* helps us reclaim simple, doable steps that reconnect us with the Earth. And as that connection deepens, we not only gain practical ways to support others; we also become more fully alive ourselves—better able to model a way of being that is meaningful, compassionate, and, yes, even fun.

Robert J. Wicks, Psy.D.

Author, *It's Good to Be Lost Once in a While* (Oxford University Press)

Professor Emeritus, Loyola University Maryland

Introduction

We begin with gratitude.

Gratitude for the land beneath our feet, the breath that moves through our bodies, and the old relationship between humans and the more-than-human world. Before we talk about theory, practice, or protocols, we start by remembering that we are nature. We have always been nature.

As I write, my mind wanders back to the mountains of southern Bavaria, where I grew up. The Alps framed my childhood like a quiet, steady presence. Weekend hikes filled with the smell of wet earth after rain, the sound of cowbells in the distance, and snow that transformed the landscape into something both wild and familiar. Nature was never a hobby or a weekend activity. It was the backdrop for every emotion of childhood: joy, loneliness, curiosity, fear. It was a companion.

As a teenager, this companionship turned into a sense of responsibility. I led a Greenpeace youth group, organizing school recycling drives and small environmental actions. I didn't yet know words like "eco-psychology" or "ecotherapy." I just knew that loving the Earth and wanting to protect it felt as natural as breathing. The line between my well-being and the well-being of the natural world was thin and permeable, even if I couldn't have explained it.

Years later, I moved to the United States to study psychology. I was deeply drawn to the complexity of the human mind and heart. Yet something felt off. Nearly everything happened indoors: therapy sessions, classrooms, clinical training. The spaces where people came to talk about their pain were often windowless rooms with humming lights and a clock on the wall. The forests, fields, and rivers that had held me as a child were nowhere in sight.

At the same time, colleagues and clients would casually say things like, "I just need to get out for a walk" or "Nature is my medicine." We knew, intuitively, that the outdoors helped us feel more grounded and alive. And yet we rarely invite nature into the healing professions in any intentional way. That gap became impossible for me to ignore.

When it came time to choose a dissertation topic, I followed that tension. I began to study the role of nature in psychological well-being, resilience, and spirituality. The data and theory were affirming, but more than that, they gave language to what my body had known since childhood: the natural world is not a luxury or an escape. It is a vital partner in human healing.

Of course, none of this begins with me—or with modern psychology. For thousands of years, Indigenous peoples across the globe have lived in a deep, reciprocal relationship with land, water, plants, and animals. Health, community, and spirituality have always been intertwined with the Earth in these traditions. As a non-Indigenous educator and practitioner, I enter this work with humility. I do not speak for Indigenous wisdom. I listen, I learn, I acknowledge, and I work to resist the patterns of extraction, erasure, and "taking without giving back" that have shaped Western mental health systems.

A guiding concept in this book is Two-Eyed Seeing, a teaching offered by Mi'kmaq Elder Albert Marshall. It invites us to look at the world with one eye grounded in Indigenous ways of knowing and the other in Western scientific understanding—and to use both eyes together. When we do this, our work becomes more whole, more accountable, and more rooted in reciprocity.

This book is written through that lens. You will not find a rigid step-by-step protocol here. Nature-Informed Therapy (NIT) is not a script. It is an orientation, a way of walking alongside people with the natural world as co-therapist, co-regulator, and companion. It applies not only to green forests or dramatic coastlines, but also to city parks, hospital courtyards, schoolyards, and small, stubborn patches of grass growing through pavement.

Whether you are a seasoned therapist, a community educator, a healthcare worker, a park ranger, a faith leader, an activist, a parent, or simply someone who feels a tug in your chest when you look at a tree, this book is for you.

In this book, Nature-Informed Therapy (NIT) refers to clinical, licensed psychotherapy that integrates nature in ethical, evidence-informed ways. Nature-Informed Care (NIC) refers to nature-based support and well-being practices that can be used in many non-clinical roles and settings. Throughout the pages ahead, I'll use NIT when I'm speaking about formal psychotherapy, and NIC when I'm speaking more broadly about care, education, and community support.

You may be:

- working in a clinical setting and longing to make your practice more embodied, relational, and connected to place;

- guiding people through grief, trauma, or transition and sensing that nature wants to be part of that sacred work;

- seeking something for yourself: a deeper sense of connection, a return to the Earth's rhythms, a breath of fresh air in your own life.

Wherever you find yourself, you are welcome here.

Throughout these pages, I use the word therapist or clinician when I am speaking specifically about formal psychotherapy: diagnosis, treatment planning, risk assessment, documentation, and adherence to professional ethics codes.

I use the words practitioner, facilitator, or helping professional when I'm speaking more broadly about people who support others in times of stress, learning, transition, and growth. This includes:

- Therapists and clinicians – counselors, psychologists, social workers, psychiatrists, and other licensed mental health professionals.

- Teachers and school staff – educators, school counselors, outdoor and environmental educators.

- Healthcare professionals – nurses, physicians, occupational and physical therapists, hospital chaplains, wellness staff.

- Parks and nature center staff – rangers, program leaders, volunteers, environmental justice advocates.

- Coaches and facilitators – life coaches, spiritual directors, youth workers, outdoor guides, retreat leaders.

- Faith and community leaders – chaplains, clergy, lay leaders, community organizers.

If you spend your days listening, teaching, supporting, or creating spaces where others can grow, you are part of the circle I am imagining as I write.

NIT and NIC share the same roots: repairing disconnection and cultivating reciprocal relationship with the more-than-human world. NIT is one branch of NIC, with additional responsibilities tied to psychotherapy: assessment, risk management, documentation, and clinical ethics.

The ROOTED™ framework you will encounter in this book is intentionally flexible. It can guide:

- a licensed therapist in a 50-minute session,
- a teacher during a 10-minute outdoor break with their class,
- a park ranger leading a grief walk,
- a chaplain sitting with someone under a hospital tree.

Because this book is meant for a wide circle of helpers, you'll see small markers throughout to show who a section or practice is primarily written for. Think of them as gentle trail signs:

- ℞ Clinician-Focused – for licensed mental health professionals. These sections address assessment, diagnosis, risk, documentation, and ethics.

- 🌿 For All Practitioners – for teachers, rangers, chaplains, coaches, and clinicians alike. These ideas and practices can be adapted in many settings without providing formal psychotherapy.

- 💭 Personal Reflection – for your own relationship with nature and your inner life, regardless of your role.

You do not have to follow these markers rigidly, but they can help you find what you need in a given moment: a clinical tool, a group activity, or simply a pause to tend to your own roots.

My hope is that you will recognize yourself and your work somewhere in these pages, and that ROOTED will give you language and structure for something you may already feel: the more-than-human world can help.

We are living through a time of sharp disconnection and, at the same time, real possibility. The world is asking us to remember. To remember that we are not separate from nature but woven into its fabric. Healing does not always begin with words. It often begins with presence, stillness, and reciprocal care.

So, we start with thanks. Thanks to the land. Thanks to our ancestors. Thanks to the teachers—human and more-than-human, who have brought us to this moment.

Let's take a breath together.

Let's begin.

Part I: Remembering (Foundations & Lens)

In Part I, we look at the roots of Nature-Informed work: how we got here, what we inherited, and why a return to relationship with the living world matters for healing.

Introduction

Chapter 1: Remembering Nature as Co-Therapist 🌿

Nature has always been part of healing, even when modern helping professions forgot how to name it. Before Nature-Informed Therapy can be understood as a model, it has to be remembered as something older and more familiar: a relationship. This chapter lays the foundation for that remembering. We begin by looking at what Nature-Informed Therapy looks like in real practice, why this work matters now, how it relates to other nature-based approaches, and what ethical responsibility asks of us as practitioners.

- what Nature-Informed Therapy (NIT) looks like in real-world practice
- why a nature-informed approach has become necessary in the current mental health landscape
- how NIT relates to and differs from other nature-based and ecotherapy models
- what it means to be an ethical Nature-Informed Practitioner, including scope and limits
- a practical exercise to begin shaping your own Outdoor Therapy Ethics Code

The Bamboo and the Storm (A Taoist Parable, Revisited) 🌿

Once there was a young man who went to seek wisdom from an old Taoist master. Heavy with worry, he asked, "Master, how can I find peace when the storms of life keep knocking me down?"

The master did not answer right away. Instead, he led the young man to a grove of bamboo. The stalks rose high above their heads, swaying and whispering to one another in the wind. Pointing to the tallest stalks, the master said, "Watch."

A strong gust barreled through the grove. The bamboo bowed low, leaves fluttering wildly, then rose again, unbroken.

"Do you see?" the master asked. "The bamboo does not resist the storm. It moves with it. Its strength lives in its willingness to bend."

They stood there in silence for a while, listening to the rustle of leaves. Finally, the master added, "Nature teaches us resilience. We cannot control the wind. We cannot stop the storms. But we can remember how to bend. When we move with life instead of hardening against it, we discover a strength that is deeper than our fear."

The young man realized something simple and important: nature is not just beautiful scenery. It is a teacher, a mirror, a quiet companion that knows something about surviving hard seasons.

Nature-Informed work is, at its heart, the choice to listen to that teacher.

In the Beginning, There Was Nature

"This is a bigger plumbing issue than we thought. We'll have to come back and dig up the yard."

It was the second major house repair in one week. As the plumber explained the estimate, my mind took off at full speed: *We just paid that other bill… What if something else breaks… What if we can't keep up…*

My smartwatch vibrated, letting me know my stress level was "elevated," as if my twitching eyelid and clenched jaw hadn't already delivered that message.

So, I did what my body has learned to do long before my mind knows what to do: I went outside.

A small trail near our home leads into a patch of woods. Within minutes, the sounds of traffic faded behind me. Birdsong and wind through branches took their place. Sunlight filtered through the leaves in shifting patterns. The air smelled faintly of earth and pine needles. My shoulders began to drop. My breath finally reached my belly instead of stopping in my chest.

The plumber's words hadn't changed. The bill hadn't disappeared. But *I* was different—more able to respond instead of spin.

Our ancestors didn't need research articles to know this: nature heals. Long before we had diagnostic codes and therapy offices, people went to forests, rivers, deserts, and mountaintops to make sense of life, to grieve, to pray, to remember who they were.

Nature-Informed Therapy and Nature-Informed Care are ways of honoring that ancient wisdom within a modern frame. It is the decision to recognize what many people already feel in their bones: healing does not belong only between four walls. Sometimes, the most honest therapy room has a dirt floor and an open sky.

What Is Nature-Informed Therapy (NIT) and Nature-Informed Care (NIC) 🌿

Simply put, Nature-Informed Therapy and Nature-Informed Care weave the healing properties of the natural world into our ways of helping.

Words matter here.

Instead of saying we *use* nature, as if the Earth were a tool in our professional toolbox, we speak of partnering with nature or collaborating with nature. We would not "use" a trusted colleague or "use" a friend. In this work, the natural world is approached with that same respect. The relationship is meant to be reciprocal. We receive support, and we are also invited to care.

We start with the understanding that disconnection from the natural world touches everyone: clients, practitioners, and the systems we work in. Like trauma-informed care recognizes that trauma is widespread and shapes how we design services, Nature-Informed work recognizes that ecological disconnection is widespread and shapes how we design everything from classrooms to clinics. Every choice we make from where we meet, to how we speak about land and place, is aimed at beginning to mend that rupture and cultivating relationship and reciprocity with the more-than-human world.

In practice, this can be as simple as walking side by side with a therapy client on a forest path while they talk about grief (NIT) or asking students to notice one small sign of resilience in a school courtyard (NIC). It might also look like sitting near a stream and letting the rhythm of the water guide a grounding breath, in with the rush, out with the pull of worry (NIT or NIC). And sometimes it's bringing biophilic design indoors: a "treehouse-like" office, natural materials, living plants, and a glimpse of sky instead of only screens (NIT and NIC).

Nature becomes more than background scenery. It participates. The wind, the temperature, the shifting light, the call of a bird all invite a more embodied, here-and-now experience. People are not only talking about their lives; they are feeling their feet on the earth as they do.

Therapy and Therapeutic: A Helpful Distinction 🌿

This book is written for two overlapping circles: licensed mental health clinicians, and the wider community of helping professionals—teachers, rangers, nurses, chaplains, community leaders, coaches, and others who support human well-being.

To keep things clear, I use two words in distinct ways:

- **Therapy** refers to clinical work provided by licensed mental health professionals within their scope of practice (assessment, diagnosis, treatment planning, risk management, documentation, and adherence to laws and ethical codes).

- **Therapeutic** refers to experiences that support nervous system regulation, connection, meaning, and growth, even outside a formal psychotherapy setting.

Many practices in this book are therapeutic in that wider sense. They can be adapted in classrooms, parks, hospitals, congregations, and community settings without becoming psychotherapy.

Nature-Informed Therapy sits squarely in the "therapy" category. Nature-Informed Care is broader: it includes many therapeutic practices that do not cross into clinical treatment.

This book is not a license to practice therapy without training. If you are not a clinician, you are still wholeheartedly invited to let the ROOTED™ model shape how you design spaces, rituals, and routines, and partner with mental health professionals when someone needs more support than your role allows.

Whenever guidance is intended specifically for clinicians (documentation, liability, crisis response), I will say so. When a practice is appropriate for a wider circle of helpers, I will name that too.

My hope is that this distinction gives you both courage and healthy limits.

What Does Nature-Informed Work Look Like? 🌿

Traditional talk therapy often focuses on thoughts, beliefs, and stories. Important work happens there. But we are not only thinking beings. We are breathing, sensing creatures with bodies that remember.

Nature-Informed work invites the whole system—mind, body, and surroundings—into the room.

Consider the difference:

- Sitting in a windowless office, fluorescent lights buzzing overhead, discussing burnout

- Walking through a park, noticing how one tree leans on another for support, talking about the same burnout

The content of the conversation might be similar, but the nervous system is receiving very different messages.

Years ago, I was presenting on anxiety treatment to a group of mental health professionals. One evaluation stayed with me. A social worker, practicing in a dense urban neighborhood, wrote that my focus on nature felt irrelevant to her clients: "There is no nature where I work."

Her comment stung. Had I failed to communicate that nature is not only forests and mountaintops?

Nature in the city might be a small community garden tucked between rowhomes, a single tree casting shade on a busy sidewalk, or a strip of sky between buildings. It might be sunlight on a client's face during a session on a bench outside the clinic, small openings that still carry real nervous-system messages.

Imagine a client with depression in the middle of a bustling city. Instead of keeping the blinds shut and the door closed, you walk together to a tiny park down the block. You pause to notice one patch of grass pushing through a crack in the pavement. You listen to a strange duet of birdsong and sirens.

After ten minutes outside, the story they tell may be the same, but their body is different: a little more grounded, a touch less numb.

This is Nature-Informed work. It does not require a dramatic wilderness expedition. It asks for a gentler decision: to let the natural world join the work.

Why Is This Work Needed?

In a remarkably short span of human history, we have become an indoor species.

Most of us spend the majority of our days under artificial lights, looking at glowing rectangles, moving from one climate-controlled box to another. Our nervous systems, however, are still tuned to something older: sunrise and sunset, cool air on our cheeks, the texture of bark, the rhythm of waves, stars at the edge of our vision.

When that relationship frays, symptoms show up:

- rising anxiety and chronic stress
- difficulty concentrating
- emotional numbness
- irritability and exhaustion that no weekend away seems to touch

Humans are out of balance, and the Earth reflects this imbalance in its own distress.

Nature-Informed Therapy and Nature-Informed Care are ways of remembering. They reconnect people to the paces and beings around us not as scenery, but as a partner in care. When therapy and helping work step outside, we are not only addressing panic attacks, grief, or trauma. We are also inviting people back into a story where they belong to a living planet, and what we belong to, we tend to protect.

Chapter 1: Remembering Nature as Co-Therapist 🌿

In 2024, I was interviewed by *The New York Times* about the growing trend of taking therapy outdoors. Another voice in the article, Dr. Petros Levounis, president of the American Psychiatric Association, expressed concern about what he saw as an informal and untested approach. His caution reflects something important. We do need clear ethics, training, and safety when taking people outside.

But when the article was published, something fascinating happened. The comment section filled quickly, not with skepticism about outdoor therapy, but with strong pushback against the idea that it was new or frivolous. Readers reminded us that humans have sought healing in nature for as long as we have stories. They wrote about walks that kept them sane, gardens that held their grief, forests that helped them stay sober.

The public recognized something that our profession is still catching up to. Bringing therapy outdoors is less an innovation and more a homecoming.

Nature's Role in Therapist Well-Being 🩺

NIT is not only good for clients. It can be a lifeline for professionals in most fields.

Many nature-informed colleagues report seeing more clients outdoors without the crushing fatigue they feel after back-to-back indoor or virtual sessions. They describe the "co-therapist" effect: nature helping to hold the intensity of the work.

> *"Integrating Nature Therapy into my private practice after recovering from intense burnout has been a giant relief for me! Taking clients outside leaves me feeling refreshed, renewed, and energized... I feel like I've got a third-party present, supporting me in my work of holding space for the client."*
>
> *— Nathalie Savell, LCPC*

> *"Whether it's through outdoor sessions, nature-based activities in my office, or simply fostering a newfound awareness of the natural world, I have seen clients better manage anxiety, depression, grief, and life changes by simply being present*

with Nature… For me, Nature is more than just a backdrop; she is an active partner in therapy."

— Gina Strauss, LCPC

When therapists are constantly indoors, we can start to feel like potted plants: technically alive, but cramped, overmanaged, starved for sunlight. Taking our work outside can rekindle:

- curiosity and creativity,

- a sense of personal meaning,

- our own connection to body and breath.

A therapist who feels more alive is better able to help clients feel alive. In this way, NIT is an act of sustainability—for us as much as for those we serve.

How Our Work Relates to Other Nature-Based Approaches 🌿

Nature-based helping practices are wonderfully diverse, and each has its place.

- Adventure therapy often focuses on challenge and accomplishments: rock climbing, ropes courses, paddling. Physical experiences become catalysts for insight and growth.

- Wilderness therapy places people in more extended, often intensive, backcountry settings, where the rigors of living outdoors shape the therapeutic container.

In both, the experience itself is often the main engine of change.

Nature-Informed Therapy and Nature-Informed Care share a love of the outdoors, but they rest on slightly different cornerstones: relationship and attachment.

In NIT, the therapeutic bond between client and clinician remains central. Nature is an active partner, but not the only one. Established modalities—cognitive-behavioral, acceptance-based, psychodynamic, somatic, humanistic, spiritually integrated—are adapted so they can unfold with the land rather than separate from it.

Many people would place this work under the broad umbrella of ecotherapy, and in many ways that would be accurate. Nature-Informed work grows from the same soil: the understanding that human well-being and ecological well-being cannot be separated. At the Center for Nature Informed Therapy, we have chosen the language Nature-Informed Therapy (NIT) and Nature-Informed Care (NIC) very intentionally. To be *informed* means to be guided by a deep understanding, to let what we know and who we are in relationship with shape the choices we make.

In NIT, what informs us is nature itself: its patterns, its rhythms, its teachings, and our ongoing relationship with the more-than-human world. We are guided by the awareness that disconnection from the natural world touches everyone, including clients, practitioners, and the systems we work within. Every aspect of NIT is shaped by this awareness and by our commitment to mending that disconnection through relationship, reciprocity, and care.

"Nature-Informed" is also language that can travel. I believe therapy should be Nature-Informed, but so should teaching, medicine, architecture, law, community organizing, and design. We can imagine Nature-Informed teachers, Nature-Informed hospital chaplains, Nature-Informed city planners. The phrase points beyond a single clinical approach toward a wider cultural shift: human professions remembering that they are nested within a living planet, not floating above it.

Nature-Based Approaches: A Quick Comparison

	Nature Informed Therapy (NIT)	Wilderness/Adventure Therapy	Forest Bathing
Main Focus	An **evidence-based, relational approach** that integrates nature into everyday mental health practices, emphasizing **building an ongoing relationship** with nature.	Use **Outdoor adventure and group bonding** to address behavioral or emotional challenges, often in multi-day wilderness settings.	A **mindful, sensory immersion** in forest environments, typically in **short guided sessions**, aimed at reducing stress and enhancing well-being.
Facilitator	**Board-certified** mental health professionals **trained in Nature Informed Therapy (NIT)**.	**Board-certified** mental health professionals specialized in adventure therapy.	Typically guided by **non-clinical** practitioners with forest-bathing training.
Targeted People	**All ages and capabilities**, from mild stress relief to deeper trauma work, in both clinical and non-clinical settings.	**Youth, young adults**, or physically capable individuals who can handle wilderness conditions and adventure activities.	Open to **all ages** with minimal physical requirements, though a certain level of mobility is helpful for walking through forested areas.
Typical Setting	**Flexible**: parks, wilderness, green spaces, indoors with nature elements, or short outdoor sessions woven into daily life.	**Wilderness** or remote outdoor environments, often for multi-day excursions or residential programs.	**Forests** or wooded areas specifically designed for a guided, immersive walk (shinrin-yoku).
Common Uses	Anxiety, PTSD, stress, grief, burnout, ADHD, individual well-being, and life satisfaction - **with nature as a constant therapeutic partner.**	Adolescent behavioral issues, juvenile delinquency, stress, PTSD, and substance abuse.	Individual well-being, physical health benefits, anxiety, stress reduction, mindfulness.
Key Difference	**Relational approach**: focuses on weaving nature into daily mental health practice, highlighting small, consistent connections over time.	**Intensive approach**: longer-term, immersive experiences in often challenging outdoor settings for behavior modification and growth.	**Immersive approach**: typically a short-term, sensory-based experience without a formal evidence-based treatment model.

Reflection for Helping Professionals 💬

Take a moment, perhaps even step outside if you can, and reflect:

- How do I personally connect with nature? Where do I feel most at ease outdoors?

- When I picture "therapy," does my mind automatically see four walls and two chairs? What happens if I imagine a trail instead?

- What small ways have I *already* invited nature into my work—perhaps without naming it?

- Where do I feel hesitant? Is it about liability, logistics, client safety, my own comfort level outdoors?

You don't need to have all the answers. Simply noticing your starting point is part of beginning the journey.

The Ethical Nature-Informed Therapist 🩺

Walking outside with a client changes the frame. Along with new possibilities come new responsibilities.

An ethical Nature-Informed Therapist:

- honors both their professional ethics code (counseling, social work, psychology, marriage and family therapy, etc.) and an emerging outdoor therapy ethic;

- seeks supervision and consultation when integrating NIT, especially in the early stages;

- considers access, inclusion, and cultural humility, and recognizes that not all clients experience nature as automatically safe;

- attends to practical matters: confidentiality on a trail, physical safety, weather, emergency plans, and boundaries around self-disclosure when you are literally walking side-by-side.

As you move through this book, I'll invite you to begin drafting your own Outdoor Therapy Ethics Code, an evolving document that weaves together:

- your professional guidelines,

- the realities of your local landscape,

- your values around caring for land, water, animals, and human community.

Think of it as a living agreement between you, your clients, and the places where you work.

Human·Nature·Healing

Code of Outdoor Care Ethics

G - R - O - W - I - N-G

G: GIVE gratitude for nature's healing capacity.

R: RESPECT local regulations, closures, and cultural protocols.

O: OWN your impact and tread lightly.

W: WORK to give back through reciprocity, service, or giving.

I: INTEGRATE ethical care across all helping professions.

N: NO harm to people, place, or practice

G: Ground your work in ongoing consultation and supervision

www.natureinformedtherapy.org

Practice: Clinician-Focused (NIT) ⚕

Thematic Nature Sessions

Purpose

To structure sessions around a theme drawn from both the client's life and the natural world, helping them experience concepts like resilience or growth not only as ideas, but as something they can see, touch, and feel.

Step 1 - Choose a Theme Together

Invite the client to choose a word or idea that fits their current season. For example:

- *Resilience* – getting back up after loss or change

- *Growth* – stretching into new roles or identities

- *Change* – navigating transition, uncertainty, or aging

Ask: "When you think of this word in your life right now, what comes to mind?"

Step 2 - Go Outside with the Theme in Mind

This can be as simple as a walk around the block, a nearby park, a campus green, or a garden in a hospital courtyard.

As you walk, invite silent observation: "Let's just look for any signs of [theme] in this place. You don't need to talk yet, just notice."

- For *resilience*: a tree that healed around a scar, a flower pushing through concrete, a bird nesting in a crooked gutter.

- For *change*: leaves in different stages, a stream shifted by heavy rain, a building mid-renovation.

Step 3 - Connect Nature's Story with the Client's Story

After a few minutes, ask:

- "What did you notice that felt like resilience/change/growth?"

- "How does that image speak to your life right now?"
- "If that tree/flower/patch of grass could talk to you, what might it say?"

Let nature take the lead in offering metaphors. You don't have to force them. They tend to arrive on their own.

Step 4 - Add a Brief Visualization (Optional)

Invite the client to close their eyes (if it feels safe) and imagine themselves as part of the scene they just observed.

For example, for growth:

"Picture yourself as a young tree in this park. Your roots are quietly thickening under the soil. Above ground, you may feel small or exposed, but below, something strong is forming. What 'roots' are you growing right now that no one can see yet?"

Step 5 - Offer a Simple "Between Sessions" Invitation

Encourage a small, doable practice:

- "Once this week, notice one living element that embodies your theme. Pause for 30 seconds and really see it."
- "If you like, jot down a sentence or make a quick sketch of what you noticed."

Over time, practices like these shift the nervous system's habit from scanning for danger to noticing support.

Practice: For All Practitioners (NIC) 🌿

Nature Scavenger Hunt (Joy, Curiosity, Resilience)

Purpose

To help participants reconnect with their everyday surroundings by noticing nature through three lenses—joy, curiosity, and resilience. This practice works well for students, staff teams, volunteers, retreat groups, or community gatherings.

Step 1: Set the Scene

Let your group know you're going to do a short outdoor activity together. You might say:

"We're going to step outside for a few minutes and look at the campus/grounds/neighborhood through three lenses: joy, curiosity, and resilience. You won't have to share anything personal—just what you notice in the natural world."

Reassure people that "nature" can be big or small: a tree by the parking lot, moss on a wall, a weed in a crack, the sky between buildings.

Step 2: Give the Simple Instructions

Invite participants to use their phone camera (or simply take mental snapshots if phones aren't available). Ask them to find three things in nature in the surrounding area:

- Something that sparks joy
- Something that evokes curiosity
- Something that signifies resilience

Encourage them not to overthink it. The first thing that speaks to them is usually enough.

Step 3: Wander and Notice

Send the group out for about 10–15 minutes. Ask them to walk in silence, just paying attention to what catches their eye.

You might add:

"Let your body slow down a little as you look around. There's no right answer. If something makes you smile, wonder, or feel a sense of 'wow, you're still here,' it counts."

Step 4: Regather and Share

Come back together in a circle or small groups. Invite each person to briefly share their three findings:

Chapter 1: Remembering Nature as Co-Therapist 🌿

- What they chose for joy, curiosity, and resilience
- Why that particular element stood out to them

Keep it optional, people can pass if they prefer to simply listen.

You can use gentle prompts like:

- "What did you notice that surprised you?"
- "Which of the three—joy, curiosity, or resilience—was easiest to find? Which was hardest?"

Step 5: Connect Back to Everyday Life

Close with a short reflection:

"These qualities—joy, curiosity, resilience—are present around us all the time, even on ordinary days or stressful evenings. Sometimes we just need a small invitation to notice."

Offer an easy ongoing practice:

"This week, see if you can spot one small moment of joy, one small spark of curiosity, or one sign of resilience in the natural world around you. You don't need to fix anything—just notice."

This way, the scavenger hunt becomes not just a one-time activity, but a gentle shift in how people move through their surroundings.

Chapter Highlights — What Matters Most 🌿

Before moving on, pause here. The central idea of this chapter is simple but far-reaching: nature is not an optional extra in human healing. It has always been part of the ground on which healing stands.

- Humans have turned to the more-than-human world for comfort, wisdom, and perspective for as long as we have stories. Nature-Informed work may look new in modern clinical practice, but its roots are ancient.
- In Nature-Informed Therapy and Nature-Informed Care, nature is not just a backdrop. The natural world participates as

a partner, inviting grounding, slowing, regulation, and meaning-making.

- Nature-Informed Therapy belongs to licensed clinicians. Nature-Informed Care belongs to a wider circle of helpers. Both aim to mend our disconnect from the natural world and deepen reciprocity.

- Skepticism from within the profession reminds us of the need for solid ethics and training. Public response has reminded us that healing in nature is anything but "novel."

- When we walk outside with people, our responsibility expands—to safety, privacy, cultural experience of nature, and to the well-being of the land itself.

Most of all, this chapter is an invitation to remember that the same Earth that holds forests, oceans, deserts, and city trees is willing to hold our grief, fear, and hope as well. We do not have to heal or help alone. We can let wind, light, soil, and the small resilient lives around us join the work.

Chapter 2: A Necessary Revolution 🌿

Many of us already turn to nature when life becomes too much. We go outside without always knowing why, and something in us settles. Yet in our professional lives, we are often taught to leave that knowing at the door. This chapter explores that split, and why so many traditional models now feel too narrow for the realities people are living through. It also asks what becomes possible when nature is no longer treated as backdrop, but as part of the healing relationship itself. We invite you to examine:

- why many traditional therapy models feel too small for the world we live in now
- what makes therapy effective
- how Nature-Informed Therapy (NIT) and Nature-Informed Care (NIC) add something essential
- NIT as a homecoming to a relationship we have nearly forgotten
- a practical way to begin integrating nature into your own work and life

The Caterpillar and the Butterfly 🌿

A caterpillar crawled along a leaf, restless with its slow progress. It wanted to move faster. From where it sat, the world seemed small: only this leaf, this branch, this narrow view.

One day it felt a tug from somewhere deep inside. Without understanding why, it spun a cocoon around itself. The safe, familiar world disappeared. There was only darkness, pressure, and the unsettling feeling that everything was dissolving.

If the caterpillar could speak, it might have said, "Something is wrong. I am falling apart."

If we were watching, we might have been tempted to encourage the cocoon to open and rescue it.

But inside that cramped, uncomfortable space, a quiet revolution was underway. Cells rearranged. Old structures broke down. A new form slowly assembled itself.

When the cocoon finally split, a butterfly emerged—fragile, trembling, wings still wet. It rested, then opened those wings to the air and discovered they could hold it. The world that once seemed so narrow now stretched in every direction.

If the butterfly could look back, it might say, "Without that time in the dark, I would never have found my wings."

Change often feels like being trapped in a cocoon. Old ways of seeing dissolve. The familiar room of our work no longer fits. It can be frightening, disorienting, and tempting to rush back to what we know. Yet nature keeps offering the same message: transformation is rarely tidy, but it is possible.

Nature-Informed Therapy and Nature-Informed Care are part of a similar transformation in our field: a slow, steady shift from doing our work in isolation from the Earth to doing our work in relationship with it.

Why We Need a Quiet Revolution

When I first started as a therapist, I was eager not only to learn theory and technique, but to find out how to stay well while sitting with so much pain.

One day, I asked a group of colleagues a simple question: "What is your number one way of taking care of yourself?" Their answers came quickly and almost in unison. One shared that they walk in the park after work, another spoke about fishing on the weekends as their reset, someone else mentioned kayaking, and another described their garden as the place where they can finally breathe. Everyone had a nature story. A favorite trail. A quiet river. A patch of soil and seeds. The more they talked, the more animated they became. You could feel bodies relax just from describing the places that supported them.

Then another realization arrived, sharp and clear: not one of us had mentioned bringing any of this into our actual work with clients.

Nature was our private refuge, our secret medicine. We trusted it to regulate our nervous systems, soften our edges, and remind us who we were. But when we stepped into the "professional" role, nature stayed outside like a loyal dog left waiting by the door.

Why?

Partly, habit. We were trained to see therapy as something that happens in offices and conference rooms, not under trees or by streams. Partly, fear. Liability, weather, logistics, stigma—there are reasons we hesitate.
And partly, the weight of tradition.

That moment with my colleagues stayed with me. If we knew, in our bones, that nature was central to our own resilience, why weren't we inviting it into our care for others?

A quiet conviction began to grow in me: Something about our models needed to change.

What We Inherited: Old Rooms, Old Maps 🌿

Modern psychotherapy is young in the long story of human healing. Many of the frameworks still taught in Western schools are only about 150 years old. Think of Freud and his contemporaries, sitting in small rooms with heavy curtains and couches, trying to make sense of dreams and symptoms. In their time, their ideas were bold and disruptive.

But the world they were working in was very different from ours. Their maps were drawn for a particular culture, era, and set of assumptions. Many were developed in urban settings, far removed from daily, working relationship with land.

Meanwhile, long before psychoanalysis or diagnostic manuals, people around the world already had rich traditions of healing. Elders, midwives, spiritual leaders, and community holders helped others navigate grief, fear, and conflict. Their "offices" were often outdoors: under trees, by rivers, around fires.

These Indigenous and land-based traditions understood something Western psychology has been slow to name: well-being is not confined to the mind. It lives in a web that includes body, spirit, community, ancestors, land, water, and sky. Healing was never just an individual project. It was relational and ecological.

As a non-Indigenous practitioner, I do not claim those traditions as my own. But I can acknowledge that many of our current models are incomplete without them. I can honor that what we now call "integrative" or "holistic" often reflects wisdom that has existed for centuries—and that was ignored or suppressed by the same systems that shaped Western mental health training.

The problem is not that Freud and others were entirely wrong. The problem is that their ideas became the *only* ideas we trusted.

When we cling too tightly to inherited models, we risk practicing therapy as if humans were brains on couches, detached from soil, seasons, and community. We keep our work inside small rooms while the larger room—the living world—waits outside.

A necessary revolution does not mean throwing everything away. It means widening the frame so that our work can take place in the world we actually inhabit.

What Actually Makes Therapy Work?

Before we can understand what Nature-Informed Therapy and Nature-Informed Care add, it helps to remember what already makes therapy effective.

Research across many different approaches has found that, regardless of technique, three elements show up again and again in successful therapy:

1. the quality of the relationship

2. the client's sense of hope and expectation

3. the actions that grow out of the work

Psychologist Bruce Wampold and others have called this the "contextual model" of psychotherapy. For our purposes, we can think of it more simply:

It is the bond, the story, and the practice that heal.

Let's look at each.

1. The Relationship: A Safe Enough Place to Be Real

At the center of good therapy is the therapeutic alliance. Clients need to feel that you are with them—not as a perfect expert, but as a steady, trustworthy companion. They need to sense that their shame, anger, confusion, or grief can be spoken without being judged or abandoned.

When we bring our work outdoors, the relationship does not stop being central. In fact, it can deepen.

Picture meeting a client in a small city park instead of a tight office. You are walking side by side. There is no direct, constant eye contact. The client can look at trees, sky, or the path beneath their feet. The nervous system is receiving messages from birdsong, moving air, changing light.

The conversation may be about betrayal, fear, or loss. But the setting is not a sealed box. It is a place where strong feelings can be held without feeling "too much."

For many people, sitting in a small office heightens anxiety. Being outdoors can lower the internal volume just enough for them to speak more freely.

2. Expectation: The Story We Tell About What Helps

The second ingredient is expectation. People need to believe that what they are doing is helpful.

When clients hear, "We know that time in nature can support your nervous system, help with mood, and deepen your sense of meaning," many feel an immediate shift. They may already have a half-remembered story about a place that helped them—a childhood tree, a lake, a grandmother's garden. The idea that those experiences are not just sentimental but genuinely therapeutic can awaken fresh hope.

This is true for Nature-Informed Therapy and Nature-Informed Care alike. When we invite people to work with us outdoors or with nature as a partner, we are not just changing the scenery. We are offering a new narrative:

"Your healing is allowed to include your relationship with the natural world. You do not have to do this alone between four walls."

For some, this feels like permission to trust what they already know. For others, it opens a new possibility they had never considered.

3. Action: New Ways of Living, Not Just Talking

The third ingredient is action. Effective therapy does not stop with insight. Over time, it helps people act differently: set boundaries, reach out, rest, grieve, play, say no, say yes.

Nature-Informed work fits here in a very practical way. It offers clients and participants small, doable experiments, such as taking a ten-minute walk in a nearby green space instead of reaching for a screen, tending a single plant as a daily act of care, pausing to feel their feet on the ground before a hard conversation, and returning to the same tree, river, or patch of sky regularly, noticing how both they and the place change over time. These are not grand gestures. They are repeated gestures. Over time, they create new grooves in the nervous system— habits of turning toward nature for regulation, perspective, and connection.

Case Story – Sarah and the Sycamore ☊

Sarah, a 32-year-old graphic designer, came to therapy after a painful breakup. She was anxious, exhausted, and deeply self-critical. On the intake form she wrote, "I feel like I'm failing at being a person."

In our first meeting, she sat rigidly in the chair, hands clenched. She apologized for "rambling" every few minutes. The room felt heavy, as if neither of us could quite exhale.

I suggested that we try something different for our next session: meeting in a nearby park. There was a wide grassy area, a few benches, and a large sycamore tree I loved. Sarah hesitated, worried that someone she knew might see her. We talked about privacy, about how we would handle greetings if that happened, and she agreed to give it a try.

The following week, we met under the sycamore. The air was cool. Children played at a distance. Leaves moved gently overhead. As we settled onto the bench, I reminded her that everything she shared remained confidential, as it would indoors.

Something shifted. The conversation unfolded more easily. At one point she said, almost surprised, "It feels less like I'm under a microscope out here."

Over several weeks, I introduced the idea of Nature-Informed work. I explained that we would continue to use evidence-based approaches for anxiety and relationships, but we would let nature join us as a partner. Together, we practiced simply observing the sycamore: its bark, the way its branches held so many small lives, the places where it had healed around old broken limbs.

Sarah started calling the tree "Olive." The name made her smile.

Her expectations began to change. She no longer saw therapy as something that happened only in her mind. She began to believe that walking, sitting, and noticing outdoors could help her nervous system settle and her sense-of-self soften.

Between sessions, she found her way to small practices—a walk around the block, a moment of stillness with her eyes lifted to the sky when the anxiety rose. They were not dramatic interventions. But they were hers.

Gradually, the room in her chest loosened. The panic no longer arrived with the same force. In its place, there was a widening space—just enough to meet what she was feeling with a little more curiosity, and a little less fear.

What we see in Sarah's story is not something new, but something remembered: the power of a trusting relationship, a sense of hope, and the quiet repetition of meaningful action. Nature did not stand apart from these. It helped them take root.

Nature's Extra Ingredient

Another client, James, came to therapy crushed by grief after his father died suddenly. He described feeling as though a heavy, gray blanket had been thrown over his life. He went to work, came home, stared at the wall, and slept. Inside our office, he could talk about the loss, but his sadness seemed to sit just out of reach, like something behind glass.

After a few sessions, I suggested a change of setting. We met at a local horticultural therapy center, a place with gardens, pathways, and a quiet stream. James was unsure. "I don't really do outdoors," he said, half joking. Still, he was curious.

We sat by the water, mostly in silence at first. Birds called in the distance. Leaves floated past on the surface of the stream.

Then a mallard duck flew in and landed almost at our feet. It stayed there, unhurried, as if it had been invited. James stared, eyes suddenly bright.

"My dad," he said, "was obsessed with ducks. He volunteered with this waterfowl group. His whole house was duck stuff—paintings, carvings, all of it. This feels… weird. Like he's saying hi."

We did not rush to interpret the moment. We let it be what it was: a point of contact, a thread between James and his father, witnessed by the living world.

Later, James told me that this experience became a turning point. For the first time since his father's death, he felt less alone. The grief was still there, but it was no longer sealed inside a small, carpeted room. It had a place in the wider world.

Traditional therapy had already offered him insight and support. But sitting by that stream, with that duck, opened something that could not be accessed through words alone.

This is one of the reasons Nature-Informed Therapy and Nature-Informed Care can be so powerful. Nature introduces an element of surprise, meaning, and shared presence that we cannot fully script. It offers what one of my colleagues calls "unscheduled grace"—moments that touch clients on a level beyond our techniques.

NIT as Homecoming 🌿

At our trainings, we greet participants with a simple phrase:

"Welcome home."

We say this because Nature-Informed Therapy is, at its heart, a homecoming return to the relationship that shaped humans across millennia.

When people step onto a forest path, a meadow, or even a campus green with others, the nervous system recognizes something familiar. This is our native habitat, the place where our ancestors learned what safety felt like, where they read the sky for weather, watched animals for cues, gathered around firelight, and told stories under stars.

In Nature-Informed Therapy, belonging emerges through a triangular attachment between client, therapist, and land.

- The **therapist** offers attuned human presence.

- The **client** brings their full humanity, their wounds, strengths, questions, and hopes.

- The **land** offers a steadier, older kind of presence: seasons, textures, rhythms that do not depend on our performance.

For clients who have experienced relational wounds, this place-based connection can be radical in its simplicity. A tree does not flinch at their anger. A river does not withdraw from their tears. A patch of ground will hold their footsteps regardless of how much they feel they "deserve" support.

Over time, many clients begin to experience a new kind of belonging—one that is not limited to a single person or office. They come home not only to the Earth, but to themselves.

How
Nature Informed Therapist
Harness Nature's Power

In Nature's Embrace, Healing Begins

1 Nature as Self-Care

Therapists rejuvenate amidst natural surroundings to maintain their occupational health.

2 Biophilic Office Design

A space that heals with elements of nature.

3 Nature Enhanced Toolbox

Integrating nature-inspired interventions for holistic healing.

4 Walk & Talk Sessions

Merging therapy and green exercise for dual benefits.

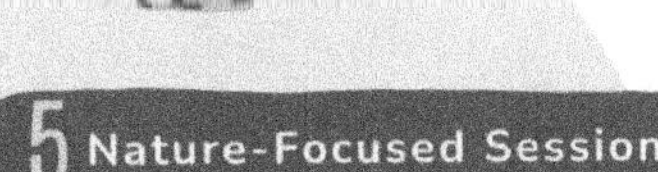

5 Nature-Focused Sessions

Crafting therapeutic experiences influenced by the natural world in both indoor and outdoor settings.

Reflection – Your Own Success Story

Take a moment to remember a time when something in your own healing or helping work really *landed*.

You might think of:

- a therapy session where you finally said something out loud that changed everything

- a student who had a breakthrough in your classroom

- a hospital visit that stayed with you for weeks

- a retreat, a conversation, or even a quiet moment alone when what had felt sealed began to open

Ask yourself:

- What made that moment possible?

- Was it the relationship, a particular practice, a story you were told?

- Did nature play any role, even in the background, a tree outside the window, a view of the sky, the feel of fresh air on your face after?

If you would like, write a short "success story" from your own life or work. Not for publication, just for you.

Then pause and wonder:

- How might nature already have been part of that story, even if you did not name it at the time?

- How could you invite nature more intentionally into your future work—or into your own self-care—to support more of those kinds of moments?

This reflection is not about judging yourself for what you "didn't do." It is about noticing how deeply connected you already are to the world around you, and how much is possible when that connection becomes conscious.

Practice: For All Practitioners (NIC) 🌿

Nature Integration Plan

Purpose

To help you intentionally weave nature into your own life and into whatever kind of care you offer—therapy, teaching, ministry, leadership, parenting, community work.

You can do this as a personal exercise or with a group of colleagues.

Step 1: Notice Your Current Relationship with Nature

Take a few minutes to reflect, or journal, on questions like:

- When do I most naturally turn to nature for comfort, energy, or clarity?
- Are there particular places—parks, porches, trees, bodies of water—that feel like "friends" to me?
- How often do I actually get to those places? Weekly? Monthly? Rarely?
- How do I usually feel after I have been outside for a while?

Write down anything that surprises you. You might realize you already have small rituals you had not named.

Step 2: Choose Three Simple Ways to Invite Nature In

Now, think about how you might become more intentional.

Choose three specific actions that feel realistic in your current season. Mix personal care and your work with others. For example:

- Outdoor journaling (personal):
  When you feel overwhelmed, step outside and sit near a tree, on a bench, or by a window with a view of the sky. Jot down what you are feeling and what you notice around you. Let the place be part of the conversation.

- Nature-Informed care moments (with others):
  If you lead groups, classes, or meetings, build in a five-minute outdoor pause: a short walk, a brief moment of quiet observation, or a simple "notice three things in nature you can see right now" prompt.

- Mindful nature walks (personal or shared):
  Take one walk each week where the only goal is to notice. Slow your pace. Pay attention to colors, textures, sounds, and scents. Let your senses lead rather than your to-do list.

If you are a clinician, one of your three actions might be:

- Nature-Informed Therapy experiment (NIT):
  Identify one client for whom an outdoor session might be appropriate. Explore logistics, safety, and consent, then try one carefully planned Nature-Informed session.

If you are not a clinician, your third action might be:

- Nature-Informed space design (NIC):
  Bring more natural elements into your everyday setting: a plant on your desk, a small bowl of stones or pinecones, nature photos or artwork, a chair positioned to face a window.

Step 3: Put It on the Calendar

Good intentions are easily swallowed by busy weeks. Choose specific days or times for each of your three actions.

For example:

- Monday: 10 minutes of barefoot standing on the grass after work

- Wednesday: brief outdoor check-in with your class or team before starting the agenda

- Saturday: one slow, device-free walk at a nearby park

Treat these as commitments to yourself and to the people you serve, not as luxuries.

Step 4: Reflect on the Impact

After each nature experience, pause for a moment of simple noticing. You can do this mentally or in writing.

Ask yourself:

- How do I feel in my body right now, compared to before?

- Did anything become a little clearer or softer?

- How does this kind of care compare to other self-care or professional tools I use?

- Is there one small thing I want to adjust next time?

Over time, these reflections will help you refine your Nature Integration Plan so that it fits the realities of your life and work, rather than adding another "should" to your list.

Chapter Highlights — What to Carry Forward 🌿

- Many helpers rely on nature for their own self-care but leave it outside their work with others. Closing this gap is part of a necessary, gentle revolution in how we think about healing.

- Traditional Western therapy models gave us important tools, but they were never meant to be the only maps. Nature-Informed work honors older, land-based traditions that have always understood healing as relational and ecological.

- Research shows that what makes therapy effective across many approaches is the relationship, the client's sense of hope, and the concrete actions that grow out of the work. Nature-Informed Therapy and Care deepen all three.

- Stories like Sarah's and James's show how nature can soften anxiety, open space for grief, and offer meaningful, unscripted moments that are difficult to reach indoors.

- Nature-Informed Therapy (NIT) names this work when it is practiced as psychotherapy by licensed clinicians. Nature-

Informed Care (NIC) names the broader orientation that can live in classrooms, parks, hospitals, congregations, and communities.

- NIT is a homecoming, restoring a triangular attachment between client, therapist, and land. For many, this place-based belonging is profoundly corrective.

- You can begin your own quiet revolution with a simple Nature Integration Plan: noticing your current relationship with nature, choosing three realistic actions, putting them on the calendar, and reflecting on what changes.

Bit by bit, these shifts help us grow our own wings—not by abandoning everything we have learned, but by letting our work be reshaped in relationship with the natural world that has been waiting, just outside the door.

Chapter 3: Eco Identity — The Story Nature Tells About Us 🌿

We all carry a story about who we are.

Some of that story is obvious: our name, our family, the language we speak. Some of it is written more subtle in the background: the smell of a river we grew up near, the way winter feels in our bones, the foods that feel like home.

This chapter is about that quieter layer, your eco-identity: the way your sense of self is braided together with landscapes, waters, seasons, and the more-than-human beings who have walked beside you.

In the pages ahead, you'll be invited to explore:

- the many threads that make up identity

- how lineage and ancestry shape our relationship with place

- how historical harms and cultural experiences can fray our connection to nature

- simple, practical ways to help yourself and those you serve rediscover their "thread to nature"

The Story of the Baobab Tree 🌿

There is an African folktale about a boy who did not know who he was.

In the center of his village stood a mighty baobab tree, thick trunk anchored deep in the earth, branches reaching wide like open arms. The elders said the baobab was a keeper of wisdom, a bridge between the living and the ancestors.

One day, the boy sat at the base of the tree and whispered, "Who am I? Where do I belong?"

The baobab answered, not with a booming voice, but with a rustle of leaves that sounded almost like speech:

"You are not just one leaf or one branch. You are part of a great tree, with roots that run deep into the earth. You are received by the stories of those who came before you. Their courage, their mistakes, their hopes—all that lives in you. To know who you are, begin by remembering where you come from."

The boy leaned his back against the rough bark and listened. As the wind moved through the branches, he felt, for the first time, that he was not alone in his questions. He was connected—downward through roots, outward through branches, backward through generations.

He returned to his village with a new kind of peace. His identity was not a puzzle he had to solve alone. It was a living story, shared with those who had walked the land before him.

Our eco-identity works much the same way. We are not just individuals moving through generic space. We are part of a living lineage, carried by places and people whose stories flow beneath the surface of our everyday lives.

Threads of Who We Are

My own story of identity began in the Alpine region of southern Bavaria.

My father taught skiing in the mountains—those long, white slopes were our winter playground. My mother had immigrated from the United States. She brought with her a very different set of stories, shaped by East Coast beaches, boardwalks, and salty air.

I grew up between these worlds: high peaks and rolling ocean, German and American, snow-packed silence and seashore chatter. I didn't have the word "bicultural" as a child, but I could feel the doubleness in my bones. My identity was not one thing. It was a woven fabric of languages, traditions, and landscapes.

Psychologists often describe identity as a sense of continuity—a feeling that, despite the changes we go through, there is a "me" that stretches across time. We knit this continuity out of memories, values, beliefs, roles, and relationships.

What we name less often is how places participate in that weaving.

Think of the first nine months of life spent floating in water. Then our very first act on Earth: a breath, an exchange with air. Long before we choose careers or labels, our bodies are in relationship with elements—water, air, temperature, light.

The landscapes we grow up in, the seasons that mark our years, and the animals we encounter or never encounter all work silently on our nervous systems and imaginations. A child raised near the ocean may come to think in tides and horizons, while a child growing up among tall city buildings may learn to read the sky in narrow slices and find beauty in small cracks of green. A child whose family works the land may carry soil under their fingernails and in their metaphors for the rest of their life.

These influences do not replace culture, race, gender, class, or other facets of identity; they intertwine with them. Take a moment to wonder how your identity is shaped by the natural environment you engage with each day, what kinds of places feel like "home" in your body, whether forest, field, city park, kitchen garden, coastline, mountains, desert, or somewhere else. And if you had to describe yourself without using any psychological terms, relying only on nature metaphors such as river, rock, storm, seed, or season, what might you choose? There are no right answers here. Only threads, waiting to be noticed.

Human Lineage: People, Places, and Memory 🌿

Lineage is more than a family tree drawn on paper. It is a web of stories, traditions, wounds, and gifts that reach across generations.

Chapter 3: Eco Identity — The Story Nature Tells About Us 🌿

When we look at our lineage through a nature-informed lens, we begin to ask:

- What places supported my people?

- What did they grow, harvest, fish, or gather?

- What landscapes shaped their fears and hopes?

Our ancestors' memories of nature—fields, forests, waters, animals—do not simply vanish. They live on in rituals, recipes, sayings, and patterns of behavior that we may carry without even knowing why.

My own journey into lineage began, somewhat uneventfully, with a DNA test.

I spat into the tube, mailed it off, and expected a tidy breakdown of percentages. What came back was a spark: strong ties to Norse and Viking groups. I found myself reading about people who spent their lives near cold seas and towering cliffs, who navigated rough waters, who told stories of mountains and forests alive with spirits.

It was not about claiming some heroic warrior identity. It was about recognizing that my love of rugged landscapes and stormy skies might have deeper roots than I realized. The more I learned, the more I noticed how often Norse stories involved talking trees, sacred groves, and a tree-of-life connecting realms. Nature was not a neutral backdrop in those tales. It was the lead character.

Your exploration may look very different. You might discover that your people were:

- farmers or shepherds

- oyster harvesters or river fishers

- miners, railroad workers, or factory laborers

- city dwellers who found their only patch of nature in a window box or weekly market

Whatever you find, your lineage can offer clues about your eco-identity:

- Did your ancestors have the privilege of access to beauty, land ownership, and leisure in nature?

- Or did they experience nature mostly as a site of hard labor, danger, or extraction?

- Were there rituals—holiday walks, seasonal foods, planting days, harvest festivals—that knit community and land together?

As practitioners, being curious about our own lineage softens us. It prepares us to approach clients' histories with respect, knowing that their relationship to nature may be shaped by forces much larger than individual preference.

When the Thread Has Been Torn

For many people, the thread back to nature has not just been forgotten, it has been damaged.

- Enslaved Africans were forced to work land they did not own, often under brutal conditions.

- Indigenous peoples were displaced from ancestral territories and punished for practicing land-based traditions.

- Immigrant laborers built railroads, harvested crops, and mined resources in dangerous circumstances with little protection.

For some descendants, nature is not automatically calming. Fields may carry echoes of exploitation. Forests may feel unsafe. Even a simple walk might be tinged with the sense that certain bodies are questioned or policed in outdoor spaces.

We cannot speak honestly about eco-identity without naming this.

When I say that every person has a thread to nature, I do not mean that every person has a pleasant history with nature. I mean something more basic and tender: we are mammals. Our nervous systems, bones, lungs, and senses are built for relationship with a living Earth.

Sometimes that relationship has been twisted into fear or avoidance. Sometimes it has been layered over with grief, anger, or numbness.

In our work, whether we are therapists or other helping professionals, we are invited to hold two truths at once:

1. Nature is a potential source of regulation, meaning, and feeling at home.

2. Nature has also been a site of harm, danger, and injustice for many.

A nature-informed approach does not romanticize the outdoors. It listens for the stories people carry—personal and ancestral—about what "outside" has meant to them.

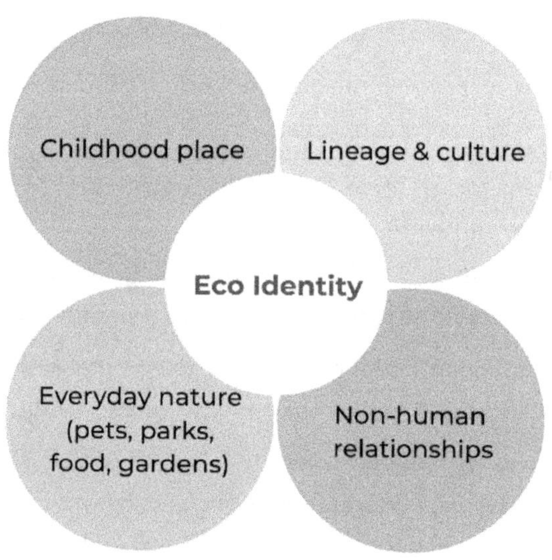

Finding the Thread in Practice – Kendrick 🩺

Kendrick, a Black man in his forties, came to therapy struggling with depression.

At our first session, he leaned back in the chair and said, "Look, I know you do this nature thing, but that's not me. I'm not into all that."

So, we didn't push it. We started where he was: a heavy numbness, a sense of moving through fog. As we talked, it became clear that much of his depression was wrapped around unresolved grief for his mother, who had died when he was a teenager.

One day, as he was describing her, a memory surfaced.

"On Saturdays," he said slowly, "she used to pack a picnic. Nothing fancy—sandwiches, chips, those canned peaches. We'd go to the park by the lake. My brothers and I would run around, but she... she just sat and watched the water. She looked... peaceful. It was the only time I remember her really resting."

As he spoke, his shoulders softened. His voice changed. You could feel the fondness and ache braided together.

I asked if he might be willing to try something: holding one session by that same lake.

He hesitated, then nodded.

We met there the following week. The bench was older, the trees a bit taller, but the water was the same. We sat in silence for a while, just listening to the gentle slap of waves against the shore.

Finally, he said, "I feel closer to her here. Like I can actually talk about her without... locking up."

We began to recognize that his thread to nature was water. Not in a generic, "water is healing" way, but in a very specific, personal way: this lake, this view, this memory of his mother's quiet rest.

So we followed that thread.

At the water's edge, grief found its own language. He began to form small rituals of remembrance. The movement of the lake reflected the movement within him. And when the sorrow rose too quickly, we returned to the body, to the feel of the bench, the sound of water, the shifting light on the surface.

Kendrick did not become "a nature guy" overnight. But the outdoors was no longer separate from his healing. The lake became a place where both his grief and his love were given room to breathe.

Finding the Thread in Practice – Laura

Laura introduced herself as "an indoor person."

"I like books, coffee shops, and my couch," she said, laughing. "Nature is… fine, I guess, but it's not really my thing."

When I asked about her family background, she mentioned that her grandparents had emigrated from Italy. As she spoke, her voice warmed. She told stories of long Sunday dinners, tomato sauce simmering for hours, herbs hanging upside down to dry in her grandmother's kitchen.

"Who grew the food?" I asked.

She paused. "You know, I never thought about it. I just remember my grandmother going out to the little garden behind the house. She'd come back with basil, oregano, all these things. She'd rub the leaves between her fingers and make us smell them."

We began to explore this more. Laura started to realize that her family's love language was deeply ecological: seeds planted, soil tended, weather watched, meals crafted from what the earth offered.

Her eco-identity was not about hiking mountains. It was about flavor, scent, and care.

As part of Nature-Informed Care, we set a small experiment: Laura would grow a few herbs on her apartment windowsill—basil, oregano, rosemary.

Over time, these small pots became more than décor. They were a living connection to her lineage. When she pinched off a leaf, she thought of her grandmother's hands. When she cooked, she felt part of a longer story of people turning earth into food, food into family, and family into memory.

Her sense of being "an indoor person" shifted. She still loved her couch and coffee shops. But now, she saw herself as someone with a thread to nature that ran through kitchens, gardens, and recipes.

Practice: For All Practitioners (NIC) 🌿/💬

Eco-Autobiography & Lineage Map

Purpose

To help you explore your own eco-identity and, by extension, to better understand and honor the nature stories your clients, students, or community members might carry.

You can do this for yourself first, then adapt it for groups or individuals you serve.

Step 1: Write Your Eco-Autobiography 💬

Set aside some quiet time. If possible, take a notebook outside or sit by a window with a view of sky, trees, or any living thing.

Write your "eco-autobiography"—the story of your life told through your relationship with nature.

Use questions like these to guide you:

- Who first introduced you to nature? A parent, grandparent, teacher, neighbor, friend, or maybe a TV show or book?
- What was your parents' relationship with nature? Did they love it, fear it, ignore it, work in it, vacation in it?
- What special nature moments do you remember from childhood?
  - A snow day?
  - A thunderstorm?
  - A backyard tree you climbed?
  - A city fountain you were obsessed with?
- What places hold meaning for you now, and what memories live there?

Let the memories come in any order. You are not writing a polished essay. You are gathering threads.

Step 2: Investigate Your Lineage

When you are ready, turn to your lineage.

You might:

- talk to older relatives

- look through old photos or letters

- recall stories about "the old country," the farm, the factory, the neighborhood

Ask:

- How did my ancestors engage with the natural world?

- Were they farmers, fishers, herders, laborers, merchants, city workers, something else?

- Did they have particular relationships with certain animals, plants, or landscapes?

- Are there recipes, songs, proverbs, or holidays that reflect those relationships?

In my case, a simple DNA test pointed me toward Norse and Viking ancestry. That opened a door to learning about people whose stories were filled with rugged coasts, thick forests, long winters, and a deep respect for the cycles of nature. It gave me new ways to understand my own draw to mountains and even inspired me to explore traditional herbs for my wellness routine.

Your discoveries may be quieter but equally meaningful. Even learning that your people lived in cramped city apartments can raise questions about what they longed for but didn't have.

Step 3: Bring It Into Daily Life

Now, choose one or two small practices that honor what you have learned.

These might look like:

- growing a single food or herb that connects you to your heritage

- visiting a park, shoreline, or trail that resembles a landscape from your family's past—or the closest version available where you live now

- learning a seasonal recipe tied to your culture and cooking it with full awareness of where the ingredients come from

- creating a simple ritual, such as lighting a candle and naming one ancestor and one place you are grateful for, before a meal or at the turn of a season

The goal is not to re-create your ancestors' lives. It is to weave a thread of continuity between you, them, and the living world you share.

A Word About Cultural Respect

As you explore, it can be tempting to adopt rituals that do not belong to your lineage—especially from Indigenous cultures whose practices have become popularized.

A nature-informed, ethical approach invites you instead to:

- start with your own people and places

- learn how *your* ancestors marked seasons, honored the land, or prayed

- create or adapt rituals that fit your life today without borrowing sacred practices from cultures that have already endured so much taking

This is not about limiting your spirituality. It is about honoring boundaries and practicing reciprocity. When we root ourselves in our own lineages and landscapes, we stand on more solid ground as we walk alongside others.

At the same time, this conversation carries important nuance. In *Restoring the Kinship Worldview*, Wahinkpe Topa (Four Arrows) and Darcia Narvaez write that many Indigenous people believe non-Indigenous teachers have no right to "teach" an Indigenous worldview, naming this concern as cultural appropriation. They note that such concern is well-grounded and must be respected in light of the massive mistreatment of Indigenous peoples over the last half millennium.

However, they also share that some Indigenous elders hold another perspective: while aware of misappropriation, they believe that with sincerity, respect, and active support for Indigenous rights, allies are necessary.

Cultural respect is not a rigid rulebook. It is an ongoing posture of discernment, reciprocity, and repair.

Chapter Highlights — What to Carry Forward

- Identity is not only psychological and social; it is also ecological. Places, seasons, and elements shape who we are just as surely as family roles and cultural labels do.

- Our lineage carries stories of how our people related to land, water, and sky. These stories can be a source of strength, grief, or both—and they deserve our attention.

- Historical injustices—slavery, colonization, forced labor—have deeply affected many people's relationships with nature. A nature-informed approach acknowledges both the healing potential of the outdoors and the harm that has taken place there.

- Every person has a "thread to nature," even if it is tangled or hidden. For Kendrick, it was water and a lakeside memory of his mother. For Laura, it was an herb-scented kitchen and family recipes.

- Writing an eco-autobiography and exploring lineage can deepen your own sense of belonging and give you language to invite clients or participants into similar explorations.

- When we root our helping work in eco-identity, we are not simply adding nature as a technique. We are recognizing that every human story is already entangled with the living world—and that healing can flow through those entanglements when we learn how to listen.

In the next chapters, we will explore how to bring this awareness into concrete practices, assessments, and interventions, so that nature is not just a backdrop to your work, but an active, honoring part of people's stories.

Chapter 4: Time Travel — Psychoevolution & Human Rewilding 🌿

Let's travel back along our species' long memory to ask why modern life often feels so mismatched to our bodies. Here we turn toward:

- how our nervous systems were shaped outdoors, in dialogue with land and other species

- the way trauma and resilience travel through generations and lineages

- how eco-separation, "place blindness," and species loneliness erode well-being

- how domestication and comfort can soften our capacity to cope

- what "human rewilding" means in everyday, realistic terms

- simple practices to rediscover your wild roots in ways that support your life and work

The Eagle Raised as a Chicken 🌿

There is an old African folktale about an eagle who forgot who she was.

A farmer found an eaglet one day and, not knowing what else to do, placed her in the chicken coop. She learned to scratch at the ground, peck at feed, and flap her wings just enough to hop onto low perches. She slept when the chickens slept, woke when they woke. She never looked up for long.

Years later, a naturalist visited the farm. He was stunned to see a young eagle living like a chicken.

"This bird was born to soar," he said. "Why does she stay on the ground?"

"She is a chicken," the farmer replied. "She has never flown. She doesn't know any different."

The naturalist disagreed. He picked up the eagle and gently lifted her high onto a fence post. "You are an eagle," he said. "Spread your wings." But the eagle looked down, saw the chickens below, and hopped back to the safety of the coop.

Again and again, the naturalist tried. Each time, the eagle returned to what she knew.

Finally, he carried her to the top of a nearby mountain, far from the coop, far from the clucking and the familiar yard. The sun was just rising. As its warmth touched her back and the wind brushed her feathers, the naturalist whispered, "Look. Remember who you are."

The eagle hesitated. She looked at the sky, then at the valley below. Something deep in her body stirred. Slowly, she spread her wings. The wind caught under them, and she lifted off. One moment of courage, and she was flying—clumsy at first, then with growing strength, circling higher and higher.

She was not learning something new. She was remembering something ancient.

In many ways, we are that eagle. We live in a world that trains us to be indoor chickens—domesticated, distracted, moving in small circles under artificial light. Nature-Informed work is an invitation to climb the metaphorical mountain, feel the sun, and ask, quietly but honestly: *What if I am more than this?*

Time Travel by Firelight

On a backpacking trip in the wooded hills of Western Maryland, I sat with a group around a crackling campfire. The air was cool. Stars began to appear, one by one, above the treetops. Someone boiled water for tea. Someone else poked at the embers, sending up tiny spirals of sparks.

I felt a wave of recognition that had nothing to do with the other people present.

It was as if I had slipped backward in time—past tents and hiking boots, past headlamps and modern gear—into a long line of humans gathered around fire. We told stories, laughed, and fell quiet in turn. The flames moved in their own language, drawing our eyes the way screens do now, but softer somehow, less demanding.

I thought of it as "caveman TV," but of course it was more than entertainment. For most of our species' existence, fires like this were central to survival. Around them, people cooked, shared food, watched over children, made tools, planned hunts, mourned the dead, and learned the stories that tied them to land and to one another.

Sitting there, I could feel how natural this setting was for my nervous system. The gentle crackle, the warmth on my face, the chorus of night insects all worked on me in a way no self-care app ever has.

Yuval Harari, in *Sapiens*, reminds us that across most of human history, our connection to nature was not just physical. It was narrative. We lived inside shared ecological stories. Seasons, rivers, animals, and weather patterns weren't background details; they were central characters in the story of who we were.

Now, most of us live under a different kind of glow.

The blue light of screens keeps us awake long after sunset. The hum of traffic replaces the chorus of night. Our "fires" are now pixels and notifications. Our eyes still gather around them, but our bodies are left wondering where the warmth went.

Nature-Informed work asks us to notice this dissonance, not as a moral failing, but as a kind of homesickness.

Psychoevolution — Our Nervous Systems Remember 🌿

Imagine stretching human history out as a long bar across a page.

Almost the entire bar would be green: millions of years lived outdoors. For more than 90% of our species' story, humans walked on varied terrain, slept under shifting skies, and woke to birdsong, wind, and the smell of earth. We hunted, gathered, fished, foraged, planted, and cared for children in direct relationship with a living world.

At the far right tip of the bar, you'd see a tiny, dense cluster: houses with HVAC systems, fluorescent-lit offices, high-rise apartments, traffic, digital devices. That little cluster is the part we call "modern life."

Our nervous systems, however, still carry the imprint of the long green stretch.

Take social anxiety. In a small tribe, being excluded didn't just feel bad. It was dangerous. To be cast out meant facing wilderness alone. Our ancestors' very survival depended on reading social cues accurately and maintaining bonds with their group. A heightened sensitivity to others' approval was adaptive.

Today, being unfollowed on social media or overlooked at a party triggers echoes of that same alarm system. Our bodies react as if a misstep could mean exile from the tribe, even if the real consequence is simply feeling awkward at the office.

Likewise, consider our need for connection and being part of. Those who survived long enough to have children were often the ones who could bond, cooperate, and share resources. Our brains are wired to scan constantly for signs of "with" or "alone."

Add to this our sensory design: natural sounds like moving water, birdsong, and rustling leaves tend to soothe the nervous system, while sudden loud noises, mechanical hums, and the constant low-level roar of traffic, ventilation systems, or alarms can agitate it. Our bodies are still tuned for a world of wind, water, and animal calls. Instead, we get alert tones, engines, and artificial humming that never fully stops.

When we recognize this, many "symptoms" begin to look less like personal failures and more like the predictable strain of an ancient nervous system trying to survive an environment it was never built for.

Nature-Informed Therapy and Nature-Informed Care begin here: not by blaming individuals for their struggles, but by acknowledging the mismatch between our psychoevolutionary design and many of our current conditions.

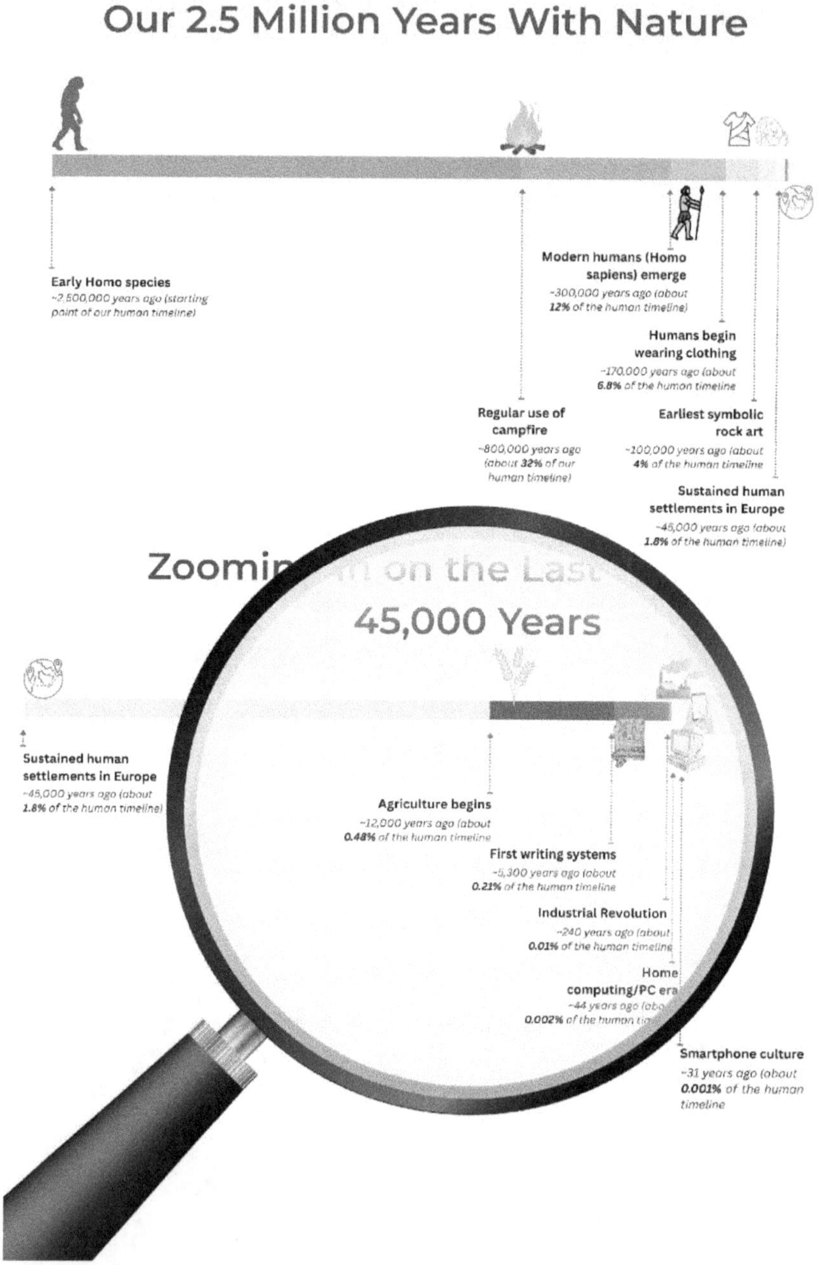

Our 2.5 Million Years With Nature

Early Homo species
~2,500,000 years ago (starting point of our human timeline)

Regular use of campfire
~800,000 years ago (about **32%** of our human timeline)

Modern humans (Homo sapiens) emerge
~300,000 years ago (about **12%** of the human timeline)

Humans begin wearing clothing
~170,000 years ago (about **6.8%** of the human timeline

Earliest symbolic rock art
~100,000 years ago (about **4%** of the human timeline

Sustained human settlements in Europe
~45,000 years ago (about **1.8%** of the human timeline)

Zooming in on the Last 45,000 Years

Sustained human settlements in Europe
~45,000 years ago (about **1.8%** of the human timeline)

Agriculture begins
~12,000 years ago (about **0.48%** of the human timeline

First writing systems
~5,300 years ago (about **0.21%** of the human timeline

Industrial Revolution
~240 years ago (about **0.01%** of the human timeline

Home computing/PC era
~44 years ago (about **0.002%** of the human timeline

Smartphone culture
~31 years ago (about **0.001%** of the human timeline

Interpreting the timeline

When we stretch human history into a single line, almost the entire bar belongs to deep time outdoors. For more than two million years, humans lived, worked, slept, and healed outside—bodies shaped by shifting light, weather, and the sounds and textures of the wider world. Our nervous systems learned in conversation with wind, soil, water, and other species. The "indoor, climate-controlled, fluorescent" chapter of our story is astonishingly new, just a tiny sliver at the bar's far edge. The magnifying glass zooms in on only the last 45,000 years—a blink beside the long green span. In that short window come agriculture, settlements, cities, writing, factories, and eventually computers and smartphones. Each step pulls us farther from the living systems we evolved alongside. Yet our bodies still expect the green world: varied terrain, weather cues, subtle sounds, and a sense of belonging to place. NIT begins here, recognizing that many modern symptoms are not personal failings but the predictable strain of an ancient nervous system trying to survive in an environment it never evolved for.

Generational Trauma and the Forest Thread 🌿

Trauma doesn't stay confined to the person who first lived through it. It moves through family systems, nervous systems, and stories.

What one generation experiences—war, displacement, addiction, emotional numbing—can become another generation's background climate. Children grow up inside adults' coping strategies. At times the most protective strategies (numbing, overcontrol, hypervigilance) also limit warmth, spontaneity, or ease.

My own family carries this pattern.

My grandfather was imprisoned in Russia during World War II. He returned home changed—a quiet man fighting battles no one talked about. My grandmother was stretched thin, raising four children in a country devastated by war. Emotional availability was a luxury they could rarely afford.

My father grew up in that atmosphere. He is a generous, gentle soul, but emotional expression was hard for him. The freeze response that helped his parents survive became part of his inherited template. Tenderness lived under layers of duty and silence.

Yet right alongside the wounds, another thread ran through our lineage: the forest.

In post-war Germany, survival meant turning to the woods. My father learned to forage mushrooms and berries, to read the land's offerings. Knowledge of wild foods created income and nourishment. Trees, moss, and soil became a second home.

When I was a child, the forest was where my father and I could meet each other most freely. He might not say much, but he would point out plants, name birds, show me where to step. The woods carried a warmth that was hard for him to access directly.

Later, as I explored my ancestry more broadly, I learned that ancient Germanic cultures held a fierce, reciprocal relationship with land. Forests were protectors. Rivers were guides. Animals were kin. Even as history layered on trauma, that older relational template with nature survived in bits and pieces.

For my family, and for many others, nature became both a witness to suffering and a bridge toward reconnection.

When we walk into a forest with clients or participants, we are not only working with the present moment. We may also be touching interrupted lineages—offering a chance to restore a sense of connection, safety, and belonging that trauma once severed.

Eco-Separation: Place Blindness & Species Loneliness 🌿

Mica Mortali, in his work on human rewilding, talks about place blindness, the way many of us have become strangers to the land we live on.

A common example: an American child who can name a hundred corporate logos but none of the trees in their own schoolyard.

It isn't that children today are less intelligent. It's that their attention has been carefully trained. Their eyes and minds have been oriented toward screens and brands, not roots and branches.

Place blindness means:

- We stop noticing the slow changes in our local environment
- We forget the names of plants, birds, and insects who share our neighborhoods
- We no longer know where our water comes from, where our waste goes, what grows naturally where we live

Alongside place blindness comes species loneliness.

For most of human history, we lived surrounded by other creatures. We watched them, listened to them, and learned from them. Our days were full of inter-species encounters: birds, insects, mammals, reptiles, fish. These relationships shaped our stories, our spiritual lives, and our sense of belonging.

Today, for many people, the only sustained relationship with another species is with a pet—and sometimes not even that. It is possible to move from home to car to office to gym and back again, spending the day almost entirely among other humans and the things humans have made.

When the nature is absent from ordinary life, something in us can grow lonely in ways we do not always know how to name. We miss the sudden pause that comes when a bird lands nearby. We miss the quiet company of a tree outside the window. We miss the grounding surprise of an animal track appeared briefly in mud.

Species loneliness rarely announces itself with drama. More often, it appears as a kind of thinning: a dullness at the edges of being, a sense that life feels flatter, narrower, less alive than it should.

Nature-Informed work can be as simple as helping people see where they already live—reversing place blindness and inviting inter-species encounters back into the story.

Domestication: When Comfort Makes Us Soft

My dog, Koda, is a wonderful illustration of domestication.

She is a dog in theory, a den animal whose ancestors roamed forests and mountains. But Koda does not like rain. She hesitates at the door if the ground looks wet, looks back at the couch, and seems to vote for the couch every time.

Wolves in Yellowstone move through storms, hunger, and cold with wild intelligence. Koda prefers soft beds, predictable meals, and dry floors. I love her exactly as she is—and she also reminds me of something about us.

We, too, have become domesticated.

Drive-thrus mean we can avoid even the short walk across a parking lot. Climate control ensures we rarely feel true cold or true heat. Lights, screens, and constant background noise shield us from darkness and silence. Our days contain fewer natural "edges."

Comfort is not the enemy. Safety and ease are real gifts. The problem is when comfort becomes our only teacher.

Resilience is forged through facing challenges and discovering that we can survive it. For most of human history, our ancestors built this resilience by adapting to the realities of the natural world—coping with shifting weather and seasons, traveling long distances on foot, tolerating periods of hunger, boredom, and uncertainty, and learning to read the landscape for signs of shelter, water, and danger. These experiences did not make life easy, but they strengthened the human capacity to endure, adapt, and continue forward.

Modern life often strips away these moments. We can move through entire days without the body ever offering a full and honest yes—without the grounded satisfaction of climbing a hill, the elemental steadiness of making a fire, or the deep quiet of a night unlit by screens.

Over time, this domestication does more than alter our habits. It can alter our self-understanding. We begin to see ourselves as fragile, even though much in us is still shaped for resilience. We forget that some discomfort can be borne, and that not all unease is harm. For some people, it is simply the edge of becoming.

Human rewilding is not a rejection of comfort. It is a remembering: that we may be sturdier, wiser, and more capable than our cushioned lives have taught us to imagine.

Human Rewilding — Answering the Call Back 🌿

Human rewilding, as described by Mica Mortali and others, is not about moving permanently into the woods or reenacting a hunter-gatherer life. It is about examining the cultural stories that keep us domesticated and reclaiming pieces of our wild nature that are still very much alive.

At its heart, rewilding asks us to reflect on where we may have become overly dependent on convenience, where we avoid discomfort that might actually strengthen us, and where we may have lost touch with our senses, our instincts, and the quiet wisdom of our own bodies.

Many forces conspire to keep us tame: family expectations, workplace culture, social media, marketing industries designed to keep us consuming. A nature-informed approach doesn't shame us for living in this world. It simply invites us to experiment with small acts of resistance.

Rewilding can take many simple forms in everyday life: saying yes to a walk in the rain instead of waiting for perfect weather, allowing yourself to feel a bit of cold or heat on your skin rather than adjusting the thermostat immediately, learning the names of the plants on your street, or sitting outdoors long enough for a bird or squirrel to forget that you are there. These are not grand gestures. They are gentle refusals to live entirely as indoor creatures. Each one strengthens the part of you that remembers: *I am an animal. I belong to Earth. I can handle more than my culture suggests.*

For many helpers—therapists, teachers, chaplains, rangers—personal rewilding also deepens professional work. When you let yourself feel more alive outdoors, you are better able to invite others into that aliveness.

Practice: For All Practitioners

Ways to Practice Human Rewilding

You do not need to try all of these. Choose one or two that fit your life right now. Let them be experiments, not obligations.

- **Sleep closer to the sky.**

  Camp, sleep under the stars, or simply nap on a porch or balcony with fresh air on your face.

- **Walk a harder path.**

  Choose a trail that challenges you a little more: hills, uneven ground. Notice how your body responds and recovers.

- **Forage with guidance.**

  Take a class or go with a knowledgeable guide to learn a few local edible plants. Let the act of finding food reconnect you to place.

- **Spend a full day outdoors.**

From breakfast to dusk, stay outside as much as realistically possible. Notice how your mood, energy, and attention shift across the day.

- **Grow something you can eat.**

  A garden is wonderful; a single pot of herbs on a windowsill counts too. Let tending and harvesting become a small ritual.

- **Declutter a bit.**

  Remove a layer of excess from your living space. See how it feels to live with less stuff and more breathing room.

- **Shrink your waste.**

  Experiment with reducing plastic use, reusing containers, or composting. Treat these as acts of relationship with land, not perfection projects.

- **Cook simply.**

  Make a meal from basic, local ingredients. Pay attention to where they came from and how they taste.

- **Go barefoot on earth.**

  Stand or walk briefly on grass, soil, or sand. Notice the sensations in your feet and legs.

- **Try the weather.**

  Safely spend time outside in rain, wind, heat, or cold. Layer appropriately, but let your body feel the elements.

- **Walk slowly, on purpose.**

  Practice mindful walking in nature, noticing each step, sound, and breath instead of rushing toward a destination.

- **Observe a cycle.**

  Watch sunrise or sunset or track one tree through a season. Let repetition teach you patience.

- **Breathe with the outdoors.**

Do your breathing exercises or meditation outside rather than inside, even if just for a few minutes.

- **Follow the sun's lead.**

  For a day or two, wake close to sunrise and wind down at sunset. Notice what changes.

- **Eat with the seasons.**

  Choose produce that is in season and grown near you when you can. Let your diet reflect the turning of the year.

- **Play with hunger.**

  If it is safe for you medically, notice the natural rise and fall of appetite between meals, rather than immediately answering every urge with food.

- **Learn one fire skill.**

  With appropriate safety and supervision, learn to start a small fire with minimal tools. Notice how it feels to coax flame from friction or spark.

- **Align with seasons.**

  Adjust your expectations of energy and productivity as the light changes across the year. Allow winter to be more inward, summer more outward.

Pick one, try it for a week, then ask: What did this awaken in me? What did it challenge? What did it change?

Practice: Personal Reflection

A Gentle Rewilding Plan (3 Tools)

You can use these three exercises for yourself or adapt them for clients, students, or groups.

1 Element Immersion

Choose one element: earth, air, fire, or water to focus on each week.

- Earth: walk barefoot, sit on the ground, touch tree bark or rocks.

- Air: notice wind on your skin, watch clouds, open a window and breathe slowly.

- Fire: safely light a candle or fire; watch the flame without multitasking.

- Water: sit by a stream, lake, or even a kitchen sink with running water; pay attention to sound and sensation.

After each immersion, jot down a few words about how you felt before, during, and after. Over time, you may discover which element your nervous system gravitates toward when you need support.

2. One-Year Wildness Inventory

Make a list of twelve nature-related activities you loved as a child or younger person. Examples:

- climbing trees

- splashing in puddles

- building forts

- watching ants on the sidewalk

- lying on the ground looking at clouds

Choose one activity each month to revisit in some age-appropriate way. Afterward, note:

- What memories surfaced?

- Did you feel silly, joyful, awkward, alive?

- What did you learn about your current relationship with wildness?

3. Rewilding Vision Board

Gather images and words that represent the kind of relationship with nature you long for.

These images might reveal landscapes that call to you, activities you feel drawn to try, or qualities you hope to embody—such as steadiness, courage, curiosity, or rest. Arrange them on a board, in a journal, or in a digital collage. Place it somewhere you'll see regularly.

This is not about creating an idealized life you must now achieve. It is about making your desire for reconnection visible, so you can move toward it one small step at a time.

Chapter Highlights — What to Carry Forward 🌿

- Our bodies and minds evolved outdoors, in close relationship with land, weather, and other species. Many modern struggles make more sense when seen as an ancient nervous system trying to adapt to a radically new environment.

- Trauma moves through generations not only as wounds but also as resilience. Lineage can carry both freeze and vitality; nature often shows up as a quiet bridge between them.

- Eco-separation—place blindness and species loneliness—leaves many of us feeling unmoored and oddly isolated, even when surrounded by people and devices.

- Domestication and comfort are not inherently bad, but when they become our only teachers, they can dull resilience and convince us we are more fragile than we truly are.

- Human rewilding is not about abandoning modern life. It is about reclaiming small, practical ways of being more fully alive, sensing, and connected to the Earth in the midst of it.

- Simple rewilding practices—walking in the rain, learning a plant's name, sitting by a fire, revisiting childhood nature joys— can restore confidence, deepen presence, and offer a richer foundation for the Nature-Informed work we do with others.

In the chapters ahead, we will build on this wild inheritance, exploring how to translate these insights into concrete assessments, interventions, and program designs that honor both our ancient roots and our very real contemporary lives.

Chapter 5: Diversity, Access & Justice — Who Gets to Feel at Home Outside? 🌿

Who gets to feel at home in nature? And who has been told, in ways loud and quiet, that they do not belong there? We'll explore:

- how identity, culture, and location shape our relationship with nature

- the impact of colonization, displacement, and segregation on land connection

- inequitable access to parks, green space, and everyday nature

- outdoor therapy as a practice of repair & reclaiming

- concrete ways to honor diversity and justice in Nature-Informed Therapy (NIT) and Nature-Informed Care (NIC)

"In nature, diversity strengthens ecosystems. The same principle applies to human organizations; the more diverse the perspectives, the more resilient and adaptable we become."

— Dr. Jane Goodall

Different Doors Into Nature 🌿

A few years ago, I sat in a community center in Baltimore with a local Black leader, listening to him talk about nature.

When people in my world say "nature," I picture hiking boots on a trail, a forest path, a backpacking trip. He pictured something very different.

"In my community," he said, "when you say 'nature,' people think about those glossy pictures—white folks in expensive gear, climbing mountains somewhere far away. That doesn't feel like us. It feels like someone else's thing."

He paused, then smiled.

"If you want to see how we connect with land, come to the park on a Sunday," he said. "You'll see grills, music, kids running around, elders watching from lawn chairs. You'll see recipes passed down through generations, herbs and seasonings our grandmothers taught us to use. That's our nature. It's right here. It smells like charcoal and chicken and collard greens. It's not hiking boots on a postcard."

His words stayed with me.

For him, and for his community, nature was not a pristine national park somewhere far away. It was the urban park where families gathered each week. It lived in food, in recipes, in shared outdoor rituals. It was embodied, cultural, and shaped by memory, survival, and belonging, not by leisure or distance from the city.

If my only image of nature is a solitary hiker standing on a remote mountain ridge, I will miss this entirely.

Nature-informed work that takes justice seriously must learn to ask different questions. Whose image of nature are we centering? Who gets left out of that image? And where is nature already present in the daily lives, memories, and practices of the people we serve?

There are many ways back into relationship with the world beyond the human-made. Some begin at a trailhead. Others begin at a picnic table, beside a portable grill, under the shade of a city tree.

Many Identities, Many Nature Stories

Nature does not meet us as blank slates.

By the time someone sits in front of you—whether in a clinic, a classroom, a chapel, or a community circle—they already carry a long list of experiences and messages about what "outside" means.

Our relationship with nature is shaped by:

- age

- race and ethnicity

- gender and sexuality

- physical and cognitive abilities

- economic class (past and present)

- geography (rural, suburban, urban, coastal, desert, mountain)

- religion and spirituality

- family stories and cultural narratives

A five-year-old's relationship with nature might be all about puddles, ants, and mud. A seventy-five-year-old might be about a particular tree they've watched for decades or a garden they can no longer tend as easily. A woman may have to think about safety and harassment in outdoor spaces in ways a man has never considered. A disabled person may experience every park through the lens of accessibility—or lack of it.

When cultural identity enters the conversation, the picture becomes even more layered. For some, land is sacred, woven into ceremony, ancestry, and kinship. For others, land carries memories of forced labor, violence, or stolen opportunity. For some communities, connection with nature unfolds through farming, fishing, foraging, or herding—ways of life that are inseparable from the land itself. For others, nature appears in quieter, everyday forms: tending balcony plants, noticing the presence of urban birds, or recognizing the familiar taste of a dish that depends on a single cherished herb. Together, these varied experiences remind us that our relationship with nature is never one-dimensional; it is shaped by history, culture, and lived experience.

When we invite nature into healing work, we are not inviting people into a neutral space. We are inviting them into a space that may carry history, comfort or threat, belonging or exclusion, joy or grief.

A nature-informed, justice-aware approach starts from humility:

"I cannot assume that nature feels the way to you that it feels to me."

History Lives in the Landscape

Our current relationships with nature did not appear out of nowhere. They sit atop long histories.

For Indigenous peoples across the globe, land is not an object or a backdrop. It is a relative, a teacher, a central part of identity, culture, and spirituality. Colonization tore at that relationship: stealing land, breaking treaties, punishing traditional practices, and rebranding stolen places as "pristine wilderness."

In what is now the United States, many national parks and protected areas were created only after Indigenous communities were pushed out. Visitors are sometimes told a story of "untouched nature," without mentioning the people who were forcibly removed so others could enjoy those views in peace.

For Black Americans and other people of color, histories of slavery, sharecropping, lynching, and Jim Crow laws have also marked the land. Fields, forests, and rivers are not always neutral settings. They can hold echoes of forced labor, racial terror, and exclusion from "recreational" spaces.

Carolyn Finney's *Black Faces, White Spaces* traces how these histories have shaped who is seen—and who sees themselves—as belonging in outdoor spaces. The dominant story of American nature has often centered white, male, able-bodied hikers, hunters, and conservationists. Everyone else is invited in only as an afterthought.

Segregation and discriminatory policies kept people of color out of certain parks, pools, and campgrounds. Violence and threats made some rural spaces unsafe. These realities echo forward. A Black family that hesitates to camp in a predominantly white rural area is not being irrational. They are remembering.

When a client or participant says, "I don't feel safe in the woods," it might be about snakes and ticks. It might also be about these deeper currents. We cannot know until we listen.

As Nature-Informed Practitioners, we are invited to:

- acknowledge these histories rather than pretending nature is apolitical

- honor that some bodies have had to move differently in outdoor spaces

- avoid assuming that what feels healing and free to us feels that way to everyone

This doesn't mean abandoning nature-based work. It means doing it with eyes open and ears tuned to the stories that live beneath the soil.

Nature Therapy as a Practice of Repair & Reclaiming 🌿/🪴

Traditional Western therapy models were developed largely by white men working indoors, in urban centers, often with white, relatively privileged clients. The room was small. The focus was on individual minds and families. The culture, land, and wider systems often remained offstage.

Practicing with cultural humility is not a slogan but a slow, ongoing process. It asks us to continually question who designed the models we use and who they were originally created to serve. It invites us to notice whose values, worldviews, and cosmologies are centered in our theories of healing, and whose have been overlooked or marginalized. With cultural humility, we intentionally make space for ways of understanding well-being and healing that emerge from outside dominant Western traditions.

Taking therapy outdoors can be part of this process—not by default, and not as a cure-all, but as an opening.

There are moments when the shift begins simply with the setting. The office, for all its usefulness, can also feel enclosed, unfamiliar, or shaped by histories that make it hard to exhale. Outside, some clients encounter a different kind of frame—one that feels less constricted, more spacious, and more their own. In these spaces, land and water and more-than-human life are not just background. They become companions in the work, offering grounding, witness, metaphor, and movement. And the outdoors can make room for forms of healing that often exceed the boundaries of a traditional session: ritual, storytelling, cultural practice, silence, and the kind of shared presence that arises when we attend together to the natural world.

Lynn Murphy's phenomenological study, *Reclaiming Healing Spaces*, explored how Black clinicians experience outdoor therapy with Black clients. Themes that emerged included:

- biophilia – a deep, often ancestral pull toward living systems

- therapeutic trust – an increased sense of safety and connection in outdoor settings

- inclusivity – a sense that nature-based work could feel more accessible and less stigmatizing than traditional office-based therapy

For some Black clients in the study, outdoor therapy reduced the sense of "going to a psychiatrist" and instead felt like "being outside with someone who gets it." Nature became a backdrop for reclaiming dignity, identity, and cultural practices that had long been marginalized in clinical spaces.

When we invite nature into therapy thoughtfully, we challenge the assumption that healing must look like two people in chairs under fluorescent lights. We make room for clients to bring their own practices: herbs, prayers, song, food, stories, into the work. And we interrupt the idea that health can only be defined by Western, white, individualistic norms.

Cultural humility is not something we "complete." It is a stance of curiosity, accountability, and willingness to change how we work as we learn.

Unequal Green: Who Gets Access to Nature?

The benefits of green space are not sprinkled evenly across our maps.

In many cities, access to green space isn't distributed evenly. Wealthier, whiter neighborhoods often have more parks, older and larger trees, cooler summer temperatures, and cleaner air and water. Meanwhile, low-income neighborhoods and communities of color are more likely to have fewer parks and smaller, less-maintained green spaces; to experience "heat islands" with higher temperatures; to live closer to highways, industrial sites, or landfills; and to face safety concerns that can make existing parks feel risky.

Even when beautiful parks exist, transportation, time off work, and the cost of gear can keep people away. A family working multiple jobs, without a car, may not be able to get to the nearest state park, no matter how inviting the trail photos look online.

In her TED Talk, environmental writer Emma Marris reflects on how our understanding of nature is often shaped by a narrow image of what "real nature" is supposed to look like—typically distant, untouched wilderness. She suggests that when we reserve our sense of awe and protection only for those places, we unintentionally diminish the value of other living landscapes around us. A vacant lot filled with wildflowers, a small city stream, or a scrappy neighborhood park can begin to seem less meaningful, even though these places are also vibrant expressions of the natural world. Marris invites us to broaden our perception and recognize that nature is present in many forms, including the everyday environments where people already live.

For many people, especially in urban settings, those "lesser" places are their only accessible nature.

Our role is not to shame anyone for not visiting remote national parks. Instead, it is to recognize the forms of nature that already exist in people's daily lives and neighborhoods. This means honoring the ways communities build relationships with the land through everyday practices—picnics, cookouts, gardening, and local rituals that bring people together outdoors. It also means advocating for greater access to nature where it is needed most, supporting efforts to plant more trees, create more parks, and ensure that safe, welcoming green spaces are available in underserved communities.

When a Baltimore community leader says, "Our nature is the park where we grill and play music," that is not a lesser version of nature. That is a true, living relationship with a place.

Nature as a human right means not only protecting mountain ranges, but also planting trees along city blocks, building wheelchair-accessible trails, and ensuring every child can safely step outside and see something green.

Therapeutic Flexibility & Empowerment

Justice-minded Nature-Informed work is flexible.

It does not assume:

- that all clients want to hike

- that all students want to sit on the ground

- that all nervous systems regulate best with quiet forest time

Instead, we see each person as the expert on their own experience and culture. We bring knowledge and ideas, but we do not dictate the "right" way to heal in nature.

Therapeutic flexibility might look like:

- asking, "What has been your favorite way of being outdoors in your life so far?" and starting there

- inviting clients to bring culturally meaningful practices into outdoor or Nature-Informed sessions—herbal teas, cooking, drumming, prayer, storytelling

- meeting in a community garden instead of a distant trail

- walking laps in a well-lit city park rather than going deep into the woods

- offering options: sit or walk; shade or sun; stillness or gentle movement

For example, a client whose cultural identity is closely tied to food and plants may feel far more at ease tending herbs in a shared garden, speaking about a grandmother's recipes, or weaving medicinal plants into rituals of self-care than sitting on a log being asked to do a traditional mindfulness exercise.

Empowerment begins with listening carefully when a client says, "This doesn't feel safe," or "This isn't my thing." It grows when we offer real choices, and when we co-create the path together rather than prescribing an experience someone is expected to follow. It also asks us to speak honestly about power, especially when differences in race, class, or culture shape the relationship between practitioner and participant.

Cultural humility is not only about whether we meet indoors or outdoors. It is about how power is accomplished and shared, whose knowledge is honored, and whose safety, comfort, and way of being are treated as essential within the healing process.

Practice: For All Practitioners 🌿/💬

The Culture Wheel: How Identity Shapes Our Nature Story

Purpose

To explore how different aspects of identity shape our relationship with nature—and to bring that awareness into your work so you can design more inclusive NIT/NIC experiences.

You can do this as a personal reflection first, then adapt it for groups, supervisees, or training cohorts.

Step 1: Draw Your Culture Wheel

On a blank page, draw a large circle and divide it into 8–10 slices, like a pizza.

Label each slice with one aspect of your identity. For example:

- Age
- Race
- Ethnicity
- Gender
- (Dis)abilities
- Sexual Identity
- Economic Class (childhood)
- Religion/Spirituality (childhood)
- Education
- Geographic Location (where you grew up)

Feel free to adjust or add categories based on what feels most relevant in your context.

Step 2: Ask Each Slice, "How Have You Shaped My Nature Story?"

For each category, jot down a few notes about how it has influenced your relationship with nature. Be as specific as you can.

Examples:

- **Age:** "As a child, nature meant playing in the alley behind our building; now in midlife, it's my morning walk before work."
- **Race/Ethnicity:** "As an Asian American woman, my parents emphasized academics over outdoor play; nature was something occasional, not central."
- **Economic Class (childhood):** "We couldn't afford camping trips, but we spent summer evenings in a free local park."

- **Geographic Location:** "Growing up near the ocean, waves and sandy feet feel like home; forests still make me a little uneasy."

- **(Dis)abilities:** "My chronic pain affects how far I can walk; accessible paths and benches make all the difference between nature feeling inviting vs off-limits."

No slice will tell the whole story. Together, they will begin to reveal a more complex picture of how you came to be who you are outdoors.

Step 3: Share (Optional) and Listen

If you are doing this in a group:

- Pair up or gather in small circles.

- Each person shares 2–3 slices that feel most important to their nature story.

- As a listener, your job is to be curious, not to compare or fix.

Notice:

- what surprises you about others' experiences

- where stories overlap

- where they differ sharply

Let diversity of experience be the teacher.

Step 4: Integrate into Practice

Now, ask yourself:

- How might my own identities make me more comfortable in some outdoor spaces and less aware of barriers others face?

- Where might I be assuming that clients or participants see nature the way I do?

- How could I adjust language, location, or activities to make my work more welcoming across identities?

Some practitioners choose to adapt the Culture Wheel as an intake tool, asking clients to share which slices feel most relevant to their nature story. Others use it in training, supervision, or team retreats to build collective awareness.

The goal is not to categorize people. It is to remember that every person's relationship with nature is shaped by many forces—and to let that awareness guide your choices.

Practice: Clinician-Focused (NIT)

An Equity Check for Your Nature-Informed Work

Use these questions as a gentle audit of your current NIT practices. You can revisit them periodically as your work evolves.

1. **Who is showing up?**

   o   Whose faces, identities, and bodies do you see most often in your Nature-Informed sessions or groups?

   o   Who is missing? What might explain that?

2. **How are locations chosen?**

   o   Are your nature spaces reachable by public transit or only by car?

   o   Are they accessible for people with mobility challenges?

   o   Do they feel culturally safe and familiar with the communities you serve?

3. **What do your images and language signal?**

   o   Do your website, flyers, and social media show only one kind of outdoor person (slim, white, young, gear-heavy), or do they reflect real diversity?

   o   Do you describe nature in ways that resonate with local communities (parks, porches, backyard gardens) or only in terms of far-off wilderness?

4. **How do you handle discomfort or fear?**

   o   When a client expresses fear or reluctance about going outdoors, do you explore the cultural and historical reasons that might be present, or do you frame it only as "avoidance to be challenged"?

5. **How much choice do clients have?**

    o   Are outdoor sessions optional? Can clients choose between different settings (indoor with plants, courtyard, park, trail)?

    o   Do you invite them to bring their own cultural practices into Nature-Informed work?

6. **How do you relate to the land itself?**

    o   Do you acknowledge Indigenous people whose land you work on?

    o   Do you model reciprocity and care for the places you use (trash, noise, impact on wildlife)?

Choose one or two areas where you feel ready to make a small, concrete change. Let justice-in-nature work be iterative rather than overwhelming.

Chapter Highlights — What to Carry Forward 🌿

- Not everyone meets nature through the same door. For some, it's a mountain trail; for others, it's a weekly barbecue in a city park, a church picnic, or a grandmother's herb garden.

- Identity—race, class, gender, ability, age, geography, culture—shapes how safe, welcome, and "normal" it feels to be outdoors. There is no one "correct" nature story.

- The legacy of our shared history continues to influence who feels at ease and welcomed in natural spaces, and who may encounter obstacles or unease there. Outdoor therapy can be a practice of cultural humility when it is done with awareness and flexibility, especially in partnership with practitioners and communities of color.

- Access to green space is deeply unequal. Justice-focused NIT/NIC work honors everyday urban nature and advocates for safe, accessible, culturally meaningful outdoor spaces for all.

- Therapeutic flexibility and empowerment mean sharing power with clients and participants: letting them define meaningful nature, choose settings, and bring their cultural practices into the work.

- Tools like the Culture Wheel and equity check-ins help practitioners examine their own lenses, reduce blind spots, and create more inclusive, responsive nature-informed offerings.

As we continue through this book, keep these questions nearby: Who is missing? Who feels welcome here? How might Nature-Informed work become not just healing, but also more just?

Part II — The ROOTED™ Mechanisms (Practice)

In Part II, we walk through the ROOTED™ mechanisms one by one, exploring how nature actually supports change in real people, real settings, and real nervous systems.

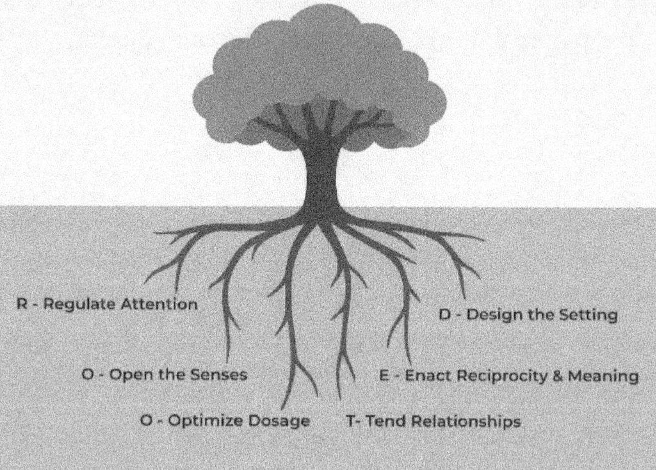

ROOTED™ Framework
Six Clinical Mechanisms of Nature Informed Therapy
R - Regulate Attention
D - Design the Setting
O - Open the Senses
E - Enact Reciprocity & Meaning
O - Optimize Dosage
T- Tend Relationships

Chapter 6: R — Regulate Attention: Feeding the Right Wolf 🌿

Attention is one of the most precious and overworked capacity we have. Many people are not failing because they are lazy or unmotivated. They are mentally frayed, over-signaled, and under-rested. Nature helps because it asks something different of the mind. It draws attention without yanking on it. In this chapter, we look at why attention breaks down, how natural settings help restore it, and what simple practices can help people come back to steadiness. Together we'll consider:

- why so many of us feel mentally frayed, even when we're "doing nothing"

- how natural environments help our brains rest and reset

- simple, short "micro-practices" that regulate attention in sessions and in everyday life

- what research tells us about ADHD, nature images, and even dorm-room windows

- a gentle practice called *Noticing Nature* that requires no extra time, only a different way of looking

The Tale of the Two Wolves — and Your Attention 🌿/💭

You may know the old teaching story about the two wolves.

A grandparent tells a child, "Inside each of us, two wolves are fighting. One is resentment, fear, greed, and self-pity. The other is kindness, courage, humility, and peace."

The child asks, "Which one wins?"

The answer: "The one you feed."

When I hear this story now, I think of attention.

Every day, countless things tug at our awareness—emails, headlines, messages, worries, memories, hopes. Some feed the wolf of fear and depletion. Others feed the wolf of steadiness and connection.

We cannot stop the wolves from existing. But we can influence which one grows stronger by how we place our attention and where we return it, again and again.

Nature-Informed work is not only about going outdoors. It is also about relearning how to aim our attention in ways that nourish rather than drain us. The "R" in ROOTED is about this: Regulate Attention.

The Weight of Attention 🌿

Not long ago, I spoke at a corporate conference for executive leaders.

The topic was productivity, but what I saw when I looked around the room was exhaustion, the kind that shows up in tight jaws, restless fidgeting, and eyes that never quite settle.

After my talk, I invited people to share what a typical day looked like.

One leader raised his hand.

"I wake up," he said, "and my first move is my phone. Email, news, maybe a few crisis messages. Coffee with more news. Then the commute, where I'm already rehearsing the day's meetings. At the office: back-to-back calls, notifications, people dropping in, endless decisions. On the way home, I'm catching up on messages. Then personal emails, social media, and maybe a show before bed. I'm always 'on,' but somehow never really focused."

Others nodded. Their stories sounded similar.

They weren't lazy. They weren't disorganized. Their attention was under siege.

Most of us live like this, to some degree. We wake to alarms, scan headlines, juggle tasks, and answer messages while stirring soup or standing in line. Our minds rarely get a true rest. Even when our bodies are still, our thoughts are sprinting.

Now imagine another kind of day.

You wake early and drive to a lake with a fishing pole. Or you sit on a park bench for an hour with no agenda. The world is quieter. Your phone stays in your pocket. Your eyes follow water, leaves, sky. You notice the rhythm of your breath.

Your responsibilities have not disappeared. But your attention has shifted from constant problem-solving to gentle noticing. Your mind is not empty, exactly, but it is less crowded. Something in you exhales.

Same brain. Different environment. Different demands on attention.

This contrast—between "office brain" and "lake brain"—is at the heart of why nature matters for the "R" in ROOTED.

Ukraine: When Slides Aren't Enough

On the first day of a Nature-Informed Therapy training with a group of Ukrainian national park rangers and botanical garden staff, the room felt heavy even before a single word was spoken.

The night before, there had been a large missile attack on Kyiv. Some participants had spent hours trying to reach family. Others had dozed in short, fitful stretches between air-raid alerts. By the time they arrived for training, their bodies were present, but their nervous systems were still braced for impact.

Before we began, one of the rangers asked, "Could we start with a moment of silence?"

Chapter 6: R — Regulate Attention: Feeding the Right Wolf 🌿

We formed a circle. Heads bowed. The silence that followed was not peaceful; it was dense. Shoulders hunched. Jaw muscles clenched. Eyes stared at the floor. It was the kind of stillness that comes not from calm, but from a collective freeze.

When the moment ended, no one reached for their notebooks. No one picked up a pen. The agenda I had carefully prepared suddenly felt irrelevant. Their minds were supposed to "take in" new material, but their bodies were still in last night's emergency.

In that moment I knew: no amount of theory would land here—not yet.

So I suggested we step outside.

We walked into the crisp morning air and gathered in small clusters beneath the trees. I gave a simple invitation:

"Turn to the people near you and share one of your favorite ways to connect with nature—something ordinary, something that helps you feel even a little more like yourself."

It wasn't a complex protocol. It was an opening to remember something familiar.

As they began to speak—about listening to birdsong at dawn, running a hand along the bark of a favorite pine, gathering mushrooms after rain—the atmosphere shifted. Faces softened. Arms uncrossed. Hands started moving as they talked. I began to hear small pockets of laughter.

The air itself felt different.

Nature did what my slides could not do. The cool breeze, the roughness of bark under fingertips, the sound of leaves moving above us—they offered something the nervous system recognizes: *You are here. You are alive. You are not alone.*

After a while, we walked back inside.

The change was subtle, but unmistakable. People were sitting a little straighter. Their eyes had more light in them. When I began to introduce the foundations of Nature-Informed Therapy, they were able to receive it. Not because I had found the perfect words, but because the land had already begun its quiet work of regulation.

That morning became an unplanned expression of the "R" in ROOTED. Attention shifted away from screens and threat and toward the living world nearby. Sensory contact with the land invited the nervous system to loosen its grip and move, however gently, out of freeze. And only then did the mind seem ready to take in something new.Sometimes, when words can't reach people, nature still can.

Attention Restoration Theory – Why Nature Helps

Psychologists Rachel and Stephen Kaplan offered a simple, powerful idea called Attention Restoration Theory (ART). It starts with a basic observation:

Our capacity to focus is limited. When we use it constantly on demanding tasks, it gets tired—just like a muscle.

The Kaplans noticed that certain environments, especially natural ones seem to give our attention a chance to rest and repair.

In lab studies, people were first given mentally taxing tasks (like unsolvable puzzles or heavy concentration work) and then shown different kinds of scenes. Those who looked at nature images (forests, lakes, meadows) recovered faster—emotionally and physiologically—than those who looked at urban scenes like buildings or busy streets. Heart rates dropped. Mood improved. Irritability decreased.

ART suggests that restorative environments share four qualities:

1. **Being Away**

   A sense of stepping out of the usual mental grind. You might still be in your city, but your mind feels like it has stepped into a different room.

2. **Soft Fascination**

Nature holds our attention—moving leaves, shifting clouds, water patterns—without grabbing it aggressively. This light grip frees up deeper parts of the mind to rest, wander, and restore.

3. **Extent**

   There is enough coherence for your mind to feel "in" the environment. This doesn't require a national park. A small park, a tree-lined street, or a courtyard with plants can provide a sense of a complete little world.

4. **Compatibility**

   The setting fits what you're trying to do. You chose it. You want to be there. You are not fighting the environment to rest or reflect.

When we offer clients and participants even brief contact with places that have these qualities, we are not just giving them a "nice break." We are helping their brains reset in ways that support therapy, learning, and daily functioning. Research by cognitive neuroscientist Marc Berman and colleagues helps illustrate this point. In one study, participants completed mentally demanding memory tasks and were then asked to take a walk either through a natural area or along busy city streets. Those who walked in the natural setting showed significantly improved attention and working memory when they returned. Interestingly, these benefits appeared even under less-than-ideal conditions, including cold Midwestern weather, strongly disliked by the participants. The findings suggest that the restorative effects of nature do not depend on perfect scenery or comfort, but on the qualities of the environment itself.

Micro-Protocols – Tiny Green Breaks for Tired Minds

The research is encouraging: even very small moments of contact with nature can help restore attention. Marc Berman and his colleagues have shown that something as simple as looking at trees through a window, viewing photographs of nature, or taking a brief "green break" can measurably support cognitive functioning.

This is welcome news for those of us who do not have a forest just outside the office door.

We do not always need long stretches of time or immersive outdoor settings to invite regulation and renewed attention. Often, what is most helpful is something brief, gentle, and repeatable, a practice that can be woven into the ordinary rhythm of a session, a class, or a conversation. These small moments may come at the beginning, as a way of arriving. They may come in the middle, when attention starts to fray or the room begins to feel heavy. Or they may come in moments when someone feels stuck, overwhelmed, or far away from themselves.

The key is not complexity, but simplicity: one object, one sense, one moment of genuine contact.

What follows are a few examples that can be adapted in your own work.

Micro-Protocol 1 – Window Watch (1–2 minutes)

Works for in-person and telehealth

"For the next minute, let's rest our attention on something outside. If there's a window, look for anything alive or moving: a branch, a cloud, a bird, the light shifting on a wall. No need to analyze it. Just watch. If there's no window, you can use a plant, a pet, or even a picture of nature on your screen. Let this be easy. When your mind drifts, simply bring it back to that one moving thing."

This engages soft fascination without demanding effort.

Micro-Protocol 2 – Three Greens, Three Breaths (1 minute)

"Look around and find three green nature things. They might be big or tiny—leaves, moss in a crack, a shirt, a photo. For each one, pause for a slow breath in and out as you look at it. Let your eyes rest there for the length of the breath."

If there is little green, you can use "three textures" (smooth, rough, soft) or "three natural colors" (brown, blue, gray).

Micro-Protocol 3 – One Object, One Sense (1–2 minutes)

Hand your client a stone, leaf, pinecone, or shell; or have them choose one where you are.

"Let's give our attention a tiny vacation. Hold this object in your hand. For the next minute, your only job is to explore it with one sense: touch or sight. Notice weight, temperature, edges, patterns, colors. If other thoughts come up, that's okay. Just return to the object."

Consider starting a small collection of natural items to build a calming nature basket for your office—to create a great variety of objects add items such as feathers, sea sponges, stones, or pieces of palo santo wood can invite touch, curiosity, and a quiet sense of connection to the natural world.

A Note on Trauma, Hypervigilance, and Dissociation

For clients with trauma histories, certain attentional practices can feel unsafe:

- Eyes closed may trigger vulnerability or flashbacks.
- Internal focus (on breath, heartbeat, or body) can be overwhelming.

For these clients:

- keep eyes open
- start with external, neutral anchors (a tree, a color, a texture)
- always offer choice: "Would you like to try this with eyes open or closed?"
- if activation spikes, quickly orient back to the room: feet on floor, sounds you can hear, objects you can name

The goal is not deep trance. It is a slight shift from scattered or flooded attention toward a steadier, more present state.

Nature and ADHD – A Different Kind of Medicine ⚕

For people living with ADHD, attention is often experienced as restless, jumpy, or easily hijacked. Many families end up in a loop of reminders, reprimands, and exhaustion.

Nature cannot replace medication or structured supports, but it can offer a powerful adjacent support.

Faber Taylor and Kuo described nature as a "safe, inexpensive, widely accessible new tool" for managing ADHD symptoms. It is not a cure-all, but it is a supportive context in which attention can organize itself with less strain.

Research has shown that:

- children with ADHD perform better on concentration tasks after time in wooded or green settings compared to more urban environments

- even a single 20-minute walk in nature can reduce ADHD symptoms and improve focus

- parents and children report more positive emotions and fewer behavior problems after green time

As clinicians and helping professionals, we can begin by widening what counts as support. We can suggest regular green breaks, not only screen breaks. We can help families notice the nearby places that feel accessible and safe—a park, a playground, a tree-lined street, even a small patch of living world woven into an ordinary neighborhood. When it fits, we can bring brief walks into the therapeutic process itself. And we can take seriously what so many parents already know in their bodies: that sometimes a child comes back from time outside feeling like a different version of themselves—softer, steadier, more available.

When we name and normalize this effect, we help families recognize nature not as an extra, but as part of the everyday fabric of care.

Photos, Windows, and Tiny Views of Green

You might be wondering: *What if my setting has almost no access to "real" nature?*

The good news: research suggests that even indirect contact can help.

- **Photographs & Images**

  Rita Berto found that people who looked at photographs of restorative environments (like forests and rivers) did better on cognitive tasks than those who saw non-restorative images. Time-limited and self-paced exposure both helped.

- **View Out the Window**

  Tennessen and Cimprich studied college students and found that those with dorm windows facing trees and nature performed better on attention tests than those whose windows faced brick walls or parking lots.

- **Green Space as Buffer**

  Van den Berg and colleagues showed that people with more green space near their homes were better protected against some of the negative health impacts of stress and difficult life events.

This doesn't mean slideshows can replace forests. But it does mean:

- hang a few nature images where people wait or work
- position chairs so that even a small patch of sky or a single tree is visible
- encourage people to take "window breaks" instead of only "coffee breaks"

Think of it this way: if the brain is thirsty, a sip of water still helps.

Practical Exercise – For All Practitioners 

Noticing Nature: A Daily Attention Reset

Dr. Holly-Ann Passmore's work on *noticing nature* shows that simply paying attention to everyday nature without changing your schedule, can lift mood and reduce anxiety. People who made a habit of noticing small natural elements in their daily routines reported more calm, connection, and gratitude.

This practice requires no extra time, only a different way of looking. You can offer it to clients, students, staff teams, or use it yourself.

Purpose

To train attention toward nearby nature in the flow of ordinary days, creating small pockets of restoration and connection without needing a special trip.

Step 1: Set a Quiet Intention

Invite participants to begin each day—or each work shift—with a simple inward commitment: *Today, I will notice at least three moments of nature.*

It can help to name how ordinary these moments may be. Nature does not have to appear as a sweeping landscape or a perfectly timed moment of awe. It may be a patch of sky between buildings, a weed pressing through a crack in the sidewalk, a bird, a tree, a houseplant, or the feel of wind, sun, or rain against the skin.

There is no need to do this perfectly. The point is not performance, but attention. Curiosity is enough.

This one lands nicely because it keeps the instructional structure while softening the list into something more invitational.

Step 2: Engage the Senses

Encourage them, as they move through the day, to pause for a few seconds whenever nature catches their eye.

Prompt:

- Sight: Notice colors, shapes, movement (light on leaves, cloud patterns, ripples in a puddle).

- Sound: Birds, wind, rain, distant thunder, even the rustle of leaves in passing.

- Smell: Damp earth, fresh-cut grass, flowers, the air after rain.

- Touch: Texture of bark, warmth of a rock in the sun, coolness of shade, the feel of air on their face.

These micro-pauses can be as short as 5–10 seconds.

Step 3: Name It Internally

Ask them to mentally note each moment with a simple phrase, such as "Noticing: the way that tree moves," "Noticing: the sound of rain on the roof," or "Noticing: the cool air when I step outside."

No journal is required (though some may enjoy one). The main thing is conscious acknowledgment. This can happen on the walk to the mailbox.

Step 4: Brief Reflection

At the end of the day—or at the end of a week—invite a quick reflection:

- *"What did I notice?"*

- *"How did these moments affect my mood or stress level?"*

- *"Did anything surprise me?"*

This can be done silently, in a few lines of writing, or shared in a group check-in.

Discussion Prompts (for Groups or Sessions)

If you are debriefing this practice with others, you might ask:

- "What were some of your favorite 'nature moments' this week?"

- "Did noticing nature shift your attention away from worry, even briefly?"

- "Did anything you noticed connect to memories, values, or spiritual beliefs?"

Over time, people often find that their mental landscape changes. The mind becomes less exclusively filled with problems and more populated with small, meaningful encounters with the living world.

Chapter Highlights — What to Carry Forward 🌿

If this chapter leaves you with anything, let it be this: attention is not a minor skill. It shapes what we notice, what we miss, and what kind of life we are able to inhabit. Nature restores attention not by demanding more effort, but by inviting a different quality of presence.

- Attention is not endless. Modern life pulls it in many directions, leaving many of us mentally tired even when we haven't done physically demanding work.

- Natural environments help restore attention by offering "being away," soft fascination, a sense of extent, and compatibility with rest and reflection. This is the essence of Attention Restoration Theory (ART).

- We do not always need deep wilderness for restoration. Brief "micro-doses" of nature—window views, plants, photos, short walks—can measurably improve focus, mood, and cognitive performance.

- For people with ADHD, time in nature has been shown to improve concentration and reduce symptoms, making it an important, accessible addition to other supports.

- Even indirect contact with nature (through images or views out a window) can support attention and buffer stress, especially when green space is present near home, school, or work.

- Micro-protocols that direct attention toward simple natural stimuli—one object, one color, one small movement—can be woven into therapy, teaching, healthcare, and daily routines.

- The *Noticing Nature* practice shows that we don't always need more time; we often need a different way of looking. By feeding our attention with small, repeated encounters with the living world, we strengthen the "wolf" of steadiness, connection, and peace.

In the ROOTED framework, Regulate Attention is not about perfect focus. It is about helping minds remember where they can rest—and inviting nature to sit beside us as we do.

Chapter 7: O — Open the Senses: Let Nature Back into the Body 🌿

So much of modern helping asks people to explain themselves before they have fully felt themselves. Our senses—smell, sound, touch, taste—are ancient pathways to regulation, memory, and meaning. By the end of this chapter, you'll be able to know:

- how scent can transport us to calmer times and places

- why natural sounds soothe a brain overloaded by noise

- what "earthing" or grounding might offer tired, inflamed bodies

- how tasting real, living food can reconnect us with land and season

- the elements—earth, water, fire, air—as simple guides for nervous-system care

- a practical exercise using smell and memory that you can adapt for many settings

The Silver Branch — A Story of Waking Up 🌿/💭

In ancient Celtic lore, there is a tale of the Silver Branch—a twig cut from a tree in Tír na nÓg, the Land of Eternal Youth.

The branch is not ordinary wood. Its leaves are silver; its blossoms are like small crystal bells. When the branch moves, the flowers chime with a sound unlike any other—part wind, part music, part something you can't quite name.

People say that whoever hears the song of the Silver Branch finds their senses coming alive. Colors grow sharper. The air feels more immediate against the skin. Ordinary things—a stone, a cup, the face of a stranger—begin to shimmer with a quiet beauty that had been there all along.

The branch does not create beauty. It reveals it.

It serves as a bridge between worlds: the everyday world of groceries and traffic, and the deeper world of wonder that we so often pass by without noticing.

I think of the Silver Branch as a symbol for what nature can awaken in us through the senses. Sometimes it is a scent that carries us back to a grandmother's kitchen. Sometimes it is the sound of rain loosening the shoulders before the mind has caught up. Sometimes it is the feel of bare feet on grass, returning us to the present more quickly than any app or instruction ever could.

When the senses wake in this way, we are not acquiring something new. We are remembering how to be fully here.

Nature-Informed Therapy (NIT) and Nature-Informed Care (NIC) can use this remembering on purpose. Instead of talking *about* experiences only in the head, we invite the nose, ears, skin, and tongue back into the conversation.

In the "O" of ROOTED, we ask:

What happens when healing begins with sensation, not explanation?

Smell — Nature's Hidden Doorway

Not long ago, I led a stress-reduction workshop for a group of professionals who were on the edge of burnout.

Their schedule for the day had been back-to-back sessions, fluorescent lights, and the kind of coffee that keeps your brain racing but never quite awake. We stepped out onto a deck overlooking a pond, a small pocket of water and trees tucked behind the conference center.

I invited everyone to stand in a loose circle, close their eyes, and take one slow breath.

Then I walked around the circle and placed a small sprig of pine into each open hand.

"Bring this gently toward your nose," I said. "No need to analyze it. Just breathe with it for a moment."

The shift was almost immediate.

Jawlines softened. Shoulders dropped. Breaths deepened. The atmosphere in the group went from tight and scattered to quiet and present.

One man opened his eyes and laughed softly. "Wow," he said. "That's my childhood camping trips right there." The scent of pine had pulled up a memory of simpler days: tents, campfires, family, nights under open sky.

We often underestimate smell. Yet of all the senses, it has one of the most direct lines to the emotional brain. A single whiff can take us back decades—before words, before analysis, straight into feeling.

In everyday life, our noses are filled with artificial smells: car exhaust, cleaning products, synthetic fragrances. Nature's scents are different. They carry the chemistry of living systems.

Dr. Qing Li's work on forest bathing has drawn attention to phytoncides, the organic compounds trees release into the air as part of their own protection against insects and disease. When we breathe them in, our bodies respond as well. Stress hormones such as cortisol begin to settle. Parasympathetic activity, the branch of the nervous system associated with rest and restoration, becomes more active. Immune function appears to strengthen. Over time, sleep and mood may improve.

None of this is entirely surprising. We did not evolve in a world of synthetic fragrance and sealed indoor air. We evolved among forests, grasslands, wetlands, and open fields. In some deep and often forgotten way, the body still knows the difference.

Aromatherapy in a nature-informed context does not have to be elaborate or clinical. It can be as simple as passing around fresh herbs such as mint, rosemary, or basil during a group, inviting someone to hold cedar or pine needles while talking through a difficult moment, or opening a window after a rainstorm and pausing together to say, "Let's breathe this in for a moment." In these small ways, scent becomes a gentle doorway back to the natural world and to the body's capacity for grounding and calm.

As you experiment, you might ask:

"Is there a natural smell that calms you or brings up good memories?"

"If you could bottle one scent from your favorite place outdoors, what would it be?"

Smell becomes a portable doorway to safety, comfort, and connection.

Sound — From Noise to Nourishment

Most of us move through a world filled with noise we barely register:

- alarms
- traffic
- HVAC systems
- the drone of machines
- constant notifications, beeps, and alerts

We tell ourselves we've "gotten used to it," but our nervous systems rarely have.

Chronic noise has been linked to increases in stress hormones, sleep disruption, anxiety, and cardiovascular strain. Studies on traffic and aircraft noise show that even when people seem to "sleep through" noise, their autonomic nervous systems do not. The body keeps reacting: tiny spikes in heart rate, blood pressure, and arousal, night after night.

In other words, the amygdala never fully gets the message that it's safe to stand down.

Now imagine trading that soundscape, even briefly, for another. People's favorite sounds in nature tend to be simple and rhythmic: the gentle tumble of a stream, leaves whispering in the wind, bird calls at dawn, the distant rumble of waves. When I ask groups, "What is your favorite sound in nature?" water almost always wins, waves lapping at the shore, rain on a roof, a creek in the woods, even a fountain in a city courtyard.

These sounds do something to us.

Research suggests that natural soundscapes—especially those shaped by water and birdsong—can help invite the nervous system into a more parasympathetic state, where the body is more able to slow down, digest, and repair. The breath deepens. Muscles begin to soften. Thoughts loosen, even if only a little, from their usual grip.

We do not need a waterfall or a wilderness retreat to work with this. Sometimes it is enough to begin simply. A recording of rain or a river can greet people as they arrive for group. In the middle of a session, we might pause and open a window, listening for whatever sounds of nature are present, however faint. And sometimes a session can close with two quiet minutes of listening—a kind of sound bath shaped not by perfection, but by whatever birds, breeze, rain, or rustling leaves are available.

You might ask:

"What sounds fill your day?"

"Which ones tire you out, and which ones nourish you?"

The goal is not perfect silence, but sound that feeds rather than frays.

Birdsong can have a surprisingly powerful effect on human well-being. Research suggests that hearing birdsong is associated with improved mood and reduced symptoms of anxiety and depression, even when the exposure is brief. In a large real-time study using smartphone reports, researchers found that people who heard birds singing reported greater mental well-being in the moment, and the positive effects often lasted for several hours afterward. Scientists believe birdsong may function as an auditory signal of a healthy, safe ecosystem, something our nervous system has evolved to interpret as reassuring and restorative. For this reason, simply pausing to notice birds in a park, garden, or neighborhood street can become a small but meaningful pathway to psychological restoration (Hammoud et al., 2022).

A teacher once shared how difficult it had become to quiet her mind. She spent her days in a lively classroom full of young children, where the noise and constant activity rarely paused. After school she would drive home to a house filled with her own four children, leaving almost no space in the day for silence. She knew she needed a way to help her nervous system settle. One day she decided to experiment during her commute. Instead of turning on the news, she played a recording of natural sounds—water flowing, birdsong, and wind moving through trees. For twenty minutes she simply listened as she drove, noticing how the sounds softened the tension she had been carrying all day. Gradually her breath deepened, her shoulders dropped, and by the time she pulled into the driveway she felt more grounded and ready to step back into the evening with her family.

Touch — Earthing and the Body's Quiet

When was the last time the soles of your feet met actual ground?

Not shoes on carpet. Skin on earth.

Many people cannot remember.

We live mostly on concrete and flooring, separated from soil, grass, and sand by rubber and plastic. The simple, mammalian act of lying on the ground or walking barefoot has become rare.

"Earthing" or "grounding" is the practice of making direct contact with the Earth through bare feet, hands, or the whole body. The idea is both poetic and literal: that touching the planet's surface can calm the body's electrical activity.

The research is still emerging, but several small studies suggest that grounding can reduce markers of inflammation and blood viscosity, improve sleep, ease certain types of pain, and lower self-reported stress and fatigue. Some experiments used grounding mats indoors. Others simply asked participants to spend time in contact with soil or grass. Early results suggest that even modest practice can help some people feel more settled in their skin.

From a NIT/NIC perspective, earthing is not a mystical cure. It is a gesture of reconnection:

- a client taking their shoes off to stand in the grass for a few minutes before a session

- a group lying on a blanket, feeling the solid support of the earth beneath their backs

- a person with chronic pain gently pressing their palms into warm sand or soil

You might invite:

"If it feels okay, try placing your hand on the ground beside you. See what you notice."

For some, the feeling will be subtle. For others, surprisingly profound.

The point is not to convince anyone of a theory, but to offer the experiment: *What happens in your body when you are in direct contact with the ground?*

During one of our programs, I invited participants to take off their shoes and socks and simply feel the grass beneath their feet. One woman in her sixties hesitated at first but eventually said yes to the invitation. A few minutes later, I noticed tears streaming down her face as she stood silently in the grass. When I asked what had moved her, she said she suddenly realized how many years she had missed this simple pleasure. She had not felt grass under her bare feet since she was a child. In that moment, the experience was not just about touching the earth, it was about reconnecting with a part of herself she had long forgotten.

Taste — Remembering Food as Earth

Taste is one of our oldest guides to safety. For most of human history, it helped us read the world with remarkable precision—helping us sense what would nourish, what might poison, what was ripe and ready, and what had already begun to spoil.

Today, the grocery aisle tells a different story.

Many highly processed foods are designed to seize the senses quickly—through salt, sugar, fat, and artificial flavor in combinations intense enough to override the body's quieter signals. Over time, the palate can be trained away from the subtle honesty of real food. Many clients describe feeling pulled toward certain snacks almost automatically, eating while scrolling, or noticing that food no longer feels satisfying, even when they are full.

Nature's foods are often less dramatic, but more truthful: a peach warmed by the sun, a handful of berries, a tomato with real flavor, bitter greens, herbs crushed between the fingers. They carry trace minerals, complex phytochemicals, and fibers the body still knows how to recognize. Research in nutritional psychiatry is increasingly pointing to links between whole, plant-forward foods and improved mood, cognition, and resilience.

From a nature-informed perspective, eating becomes more than calorie management. It becomes a way of coming back into relationship. We begin to feel the seasons in the body—spring greens, summer fruits, autumn roots. We remember that food does not begin on a shelf, but in soil, rain, sunlight, and labor. And through sensory attention, taste can become a place of gratitude, mindfulness, and re-entry into the present moment.

This conversation does not require an overhaul. It can begin gently. Sometimes it starts with slowing down over a single blueberry or a slice of apple, noticing texture, temperature, and the way flavor changes over time. Sometimes it begins with a simple question: *What is one food that reminds you of home, and why?* And sometimes it opens through shared story—inviting people to bring a small piece of what I think of as "nature food," perhaps a fruit, a nut, or an herb, and to speak not only of its taste, but of what it carries.

Taste, like smell, can become a direct path to memory, meaning, and a place to return to.

One season, I made a small personal commitment: to join a local orchard and pick my own fruit throughout the growing season. It began with strawberries in early summer, bright red and warm from the sun. A few weeks later came blueberries, then peaches and apples as the air slowly shifted toward autumn. By the end of the season, I was walking through the pumpkin patch, choosing the last harvest of the year. Every few weeks the land offered a different gift, and the rhythm of returning again and again reminded me that nature does not give everything at once—it unfolds slowly, season by season, inviting us to come back and receive what is ready.

The Language of the Elements

Across cultures, people have turned to the elements—earth, water, fire, air—to make sense of the inner life.

In NIT/NIC, we are not using the elements as rigid metaphysics, but as simple, sensory metaphors for nervous-system states and needs:

- Earth – stability, ground, structure
- Water – emotion, flow, flexibility
- Fire – energy, transformation, passion
- Air – breath, movement, perspective

This language is often easier for clients than clinical terms. Someone might say:

- "I feel ungrounded" (earth)
- "I'm flooded" (water)
- "I'm burning up with anger" (fire)

- "I can't catch my breath" (air)

We can then respond not just with words, but with elemental practices:

- Earth: holding a stone, leaning against a tree, lying on the ground

- Water: watching a stream, letting something float away, drinking slowly

- Fire: writing something down and safely burning it, sitting by a campfire

- Air: breathwork outdoors, watching clouds, walking in the wind

Recently, our team led an "anger release fire."

Participants were invited to bring whatever they were ready to acknowledge: resentments, disappointments, old hurts that had gone unspoken. We gathered around a safely contained fire and, when each person felt ready, invited them to feed the flames with what they wished to release. Sometimes this took the form of written words. Sometimes it was as simple as naming what had been carried and placing a small stick into the fire.

The task of the group was not to fix or interpret. It was to witness.

What might otherwise have leaked sideways into daily life—through sharpness toward a loved one, or a quiet cruelty turned inward—was given a place to move. In this way, fire became not a symbol of destruction, but of transformation.

Elemental language can help keep this kind of work grounded and accessible. It offers an entry point even for those who are wary of anything that feels overtly spiritual. It links emotional states to tangible acts: holding a stone, watching water, breathing with the wind. And it gives us a shared vocabulary that often feels less clinical and more deeply human.

At times, a simple question is enough to open the way: *If your nervous system were an element today, which would it be?* Or, *What element do you most need right now—more earth, more water, more fire, or more air?*

From there, the next small step can begin to reveal itself.

Practice: For All Practitioners 🌿/💬

Smell & Memory — A Gentle Path to Safety

Purpose

To use natural scent as a bridge to positive or comforting memories, supporting regulation and emotional connection.

You can offer this as an individual exercise, in groups, or as a personal reflection.

Step 1: Gather Simple Scents

Choose a few natural, accessible scents such as:

- pine or cedar sprigs
- fresh herbs (mint, rosemary, basil, thyme)
- citrus peel (orange, lemon)
- a handful of soil in a small dish
- dried lavender or other flowers

If you are indoors without access to fresh materials, use high-quality essential oils or dried herbs/spices from a kitchen.

Arrange them in small jars or bowls.

Step 2 – Set the Tone

Invite participants to sit comfortably. You might say:

"We're going to explore how smell and memory are connected. There's no right or wrong experience here. If at any point a scent feels uncomfortable, you're always free to pass or choose another."

Emphasize choice, especially for people with trauma histories.

Step 3: Smell, Then Pause

Offer one scent at a time.

Prompt:

"Bring this gently toward your nose. Take a slow breath in, then out. Don't worry about identifying it. Just notice your body's reaction. What happens in your jaw, your shoulders, your chest?"

Allow a few seconds of quiet after each inhale.

Step 4: Invite Memory

After a moment, ask:

- "Does this smell remind you of anything or anyone?"
- "Where, if anywhere, does this take you?"

Encourage participants to let **pleasant or neutral** memories lead. If uncomfortable memories arise, validate them and invite a different scent. You can offer gentle prompts:

- "A place?"
- "A season?"
- "A person?"
- "A time of life?"

Step 5: Reflect (Briefly)

If appropriate, ask participants to jot down a few lines:

- Where were you?
- Who were you with?
- What were you feeling then?
- What do you notice in your body remembering it now?

In a group, invite those who feel comfortable to share one scent + one memory. Ask listeners to simply receive, not interpret.

Step 6: Integrate

Close by naming aloud what this exercise shows:

- our senses are powerful gateways to memory and emotion
- nature's fragrances can offer quick access to regulation and comfort

- when life feels overwhelming, something as simple as smelling pine, soil, or herbs can be a way home to ourselves

Encourage participants to choose one scent they might keep near their workspace, bedside, or in a small vial in their bag as a portable anchor.

Chapter Highlights — What to Carry Forward 🌿

- Our senses are ancient pathways to regulation and connection. Nature doesn't just calm the mind through ideas; it reaches us through smell, sound, touch, and taste.

- Natural fragrances—like pine, cedar, damp soil, and rain—can lower stress hormones, support immune function, and quickly shift mood, in part through compounds like phytoncides released by trees.

- Soundscapes matter. Chronic noise wears down the nervous system, while water sounds, birdsong, and wind often invite the parasympathetic system to come forward, offering rest and repair.

- Earthing or grounding—simple skin contact with soil, grass, sand, or water—may help reduce inflammation, support sleep, and ease stress, while offering a felt sense of being literally supported by the earth.

- Food is one of the most direct ways we meet nature. Whole, minimally processed foods reconnect us with seasons and place, supporting physical and mental well-being in ways synthetic flavors cannot.

- Elemental language (earth, water, fire, air) gives clients and practitioners a simple, accessible way to describe inner states and choose concrete, sensory practices that match what they need.

- Practices like Smell & Memory help people locate small pockets of safety and comfort inside their own histories. A single scent can be a bridge: from office to forest, from overwhelm to a remembered moment of peace.

As you move forward in the ROOTED framework, remember: Open the senses is not an extra step; it is the doorway. When we let the body participate in healing, nature has many more ways to reach us.

Chapter 8: O — Optimize Dosage: How Much Nature Is Enough? 🌿

In this chapter, we turn toward a practical question that shapes all of our Nature-Informed work:

How much nature is enough, and what kind?

Together we'll try to understand:

- why "all-or-nothing" thinking about nature keeps many of us stuck

- the *Three-Day Effect* and what happens to the brain in longer immersions

- the "Wilderness Effect" and why civilization seems only "four days deep"

- how to think in terms of nature doses rather than ideal getaways

- the *Nature Pyramid* as a guide for daily, weekly, seasonal contact

- a simple practice to build your own nature-informed self-care plan

"Wilderness is not a luxury but a necessity of the human spirit. And as civilization continues to advance, it becomes more and more important to have that wilderness in our lives."

— Edward Abbey, Desert Solitaire

The "Someday Retreat" Trap 🌿/💭

For years, I carried a story in my head that sounded something like this:

Chapter 8: O — Optimize Dosage: How Much Nature Is **Enough?**

"If I could just get away for a long weekend—somewhere wild, with no Wi-Fi—that would fix it. I'd finally rest. I'd finally reset. I'd finally feel like myself again."

Maybe you know this story too.

In my imagination, the cure for my stress was always a big experience—three days in the mountains, a week in the desert, or a silent retreat by the ocean.

The problem was not that these ideas were bad. The problem was that they were rare. Life stayed full—clients, emails, family, deadlines. The magical window for a perfect retreat never quite opened.

So, I did what many of us do: I waited for the big thing and, in the meantime, did very little.

My nervous system, however, did not wait. It accumulated tension, fatigue, and a growing sense of disconnection—from nature, from other people, and from myself.

At some point, I realized:

It's not just that I *don't have enough time* for nature. It's that I've been telling myself a story that nature only "counts" in big doses.

What if that story is simply wrong?

Yes, multi-day trips can be powerful. We'll talk about those. But the truth is, most of the healing nature offers comes not from dramatic escapes but from small, steady doses—tiny contacts that accumulate like water filling a cup.

Optimizing dosage is about learning to work with the lives we actually have, not some future fantasy schedule.

The Three-Day Effect — Cleaning the Mental Windshield 🌿/⚕

David Strayer, a cognitive psychologist at the University of Utah, explored the profound impact of nature on our mental faculties through what he coined the "Three-Day Effect."

In one study, he took participants on a three-day backpacking trip. No phones. No laptops. Just people, packs, and a landscape rich with trees, rocks, water, and open sky.

After three days, their performance on creative problem-solving tasks improved dramatically—often by 40–50%. People described feeling clearer, more focused, and more alive. It was as if someone had washed their mental windshield.

Strayer's work points us toward something many of us already know in our bones: the mind was not made for constant multitasking, perpetual alerts, and an endless stream of digital input. When we step away from screens and re-enter the living world, attention begins to loosen, rest, and reorganize. Even a relatively brief period of immersion—three days, in his research—can bring a noticeable shift.

What matters, though, is not only the number of hours. It is also the quality of our attention. Are we truly unplugged, or still half-reaching for messages and updates? Are we allowing the landscape to set the pace, or are we dragging our usual urgency into the trees and calling it rest?

What I take from Strayer's findings is both simple and hopeful: a few days in nature can support cognitive renewal. Creativity and problem-solving begin to open again. Mental clarity returns as some of the inner fog starts to lift. Technology matters here too; part of what restores us is not only the presence of the natural world, but the absence of constant interruption.

There is something deeply encouraging in this. A long weekend outdoors, something that may seem modest by modern standards, can still help the mind remember itself. Extended immersion has a way of loosening what daily life so often tightens.

The Wilderness Effect — "Civilization Is Only Four Days Deep"

If three days in nature can begin to clear the mind, what becomes possible when we stay longer?

Ecopsychologist Rob Greenway, drawing from many wilderness treks, described what he called the *Wilderness Effect.* After guiding groups through the first days of discomfort, boredom, and predictable resistance—no cell signal, no shower, no familiar conveniences—he began to notice a pattern.

Around day four, something often shifted.

People began to feel more fully here. More real. More present to themselves and to one another. A quiet sense of expansion would emerge, along with a rekindled aliveness that daily life had kept buried beneath noise and obligation. Greenway's conclusion was both simple and striking: civilization is only four days deep.

It often takes that long for the first layer of modern habit to begin to loosen—the reflex to check a device, the low hum of schedules and social demands, the restless scanning for what comes next. By then, the nervous system has started to entrain to another rhythm entirely: sunrise and sunset, hunger and fatigue, the pace of walking, resting, cooking, and watching the sky.

Longer immersion often deepens the effect. People return with a different kind of clarity, a softened or steadier emotional life, and a renewed sense of belonging on the Earth. What once felt urgent may no longer seem as important. What truly matters has a way of coming back into view.

The Wilderness Effect reminds us that some forms of healing ask more of time. Timothy Beatley's Nature Pyramid offers a helpful frame here, placing wilderness immersion at the summit: powerful when available, but not required for the everyday nourishment of a nature-connected life.

When I lead weeklong backpacking trips, I often witness this shift directly. The first days on the trail are usually the hardest. Bodies ache. Packs feel impossibly heavy. The miles can seem endless. People are tired, and sometimes wondering why they agreed to come. It is as if the body is brushing up against something ancient it has not been asked to remember in a long time—our inheritance of walking, carrying, resting, and living in close relationship with the natural world.

Then, somewhere around the third day, the trail begins to feel different. Bodies start to adapt to the rhythm of movement, daylight, and rest. Sleep comes more easily. Appetite returns. A clean, honest tiredness replaces the more jagged exhaustion people brought with them.

By the fourth day, the conversations often change as well. They deepen. With the usual distractions stripped away, people begin speaking more openly about their lives. They listen differently. They connect in ways that feel less performative and more true. Something in them has settled enough to let what is real come forward.

This is part of what extended time outdoors can offer: not just relief, but reorientation.

Immersion in Nature — Letting Nature Saturate the Senses 🌿

We can think of immersion as time when nature is not a side dish but the main course. It might be a multi-day backpacking trip, a guided forest-therapy retreat, canoe or kayak camping along a river, a stay at a nature lodge or eco-retreat, a meditation retreat in a quiet natural setting, or even a personal "sit spot" retreat—returning to the same place for several days in a row.

What these experiences share is a high ratio of natural to human-made stimuli, fewer interruptions from devices and urban noise, and enough time for the senses to fully engage. Research on ecotherapy and nature immersion suggests that longer exposures can reduce anxiety and depressive symptoms, calm chronic stress responses, deepen feelings of connection and meaning, and support reflection—especially during grief or big life decisions. In *Heartbreak: A Personal and Scientific Journey*, Florence Williams describes how walking—often for long stretches outdoors—became a simple but powerful way to tend to the grief of her divorce and slowly find her footing again.

Immersion is not always dramatic. Often it's simply staying in one place long enough for the forest, lake, or field to stop feeling like "scenery" and start feeling like company.

But if immersion sits near the top of the dosage scale, what belongs at the base?

Immersion is not always dramatic. often it is simply staying in one place long enough for the forest, lake, or field to stop feeling like "scenery" and start feeling like company.

Nature Pyramid — A "Balanced Diet" of Green

Are there "minimum daily requirements" for nature, similar to nutritional guidelines for physical health?

Since 2012, Tim Beatley and his colleague Tanya Denkla-Cobb at the University of Virginia have pioneered the idea of a "Nature Pyramid" to help us think about varying degrees of nature exposure and its different effects. The Nature Pyramid is inspired by the well-known food pyramid and illustrates recommended doses of nature across different intensities and frequencies, from brief, everyday encounters to immersive wilderness experiences. This model highlights that while a long hike in the woods may be valuable, shorter daily interactions with nearby nature also contribute to our health and well-being.

Layers of the Nature Pyramid

Inspired by the food pyramid, it suggests that we need frequent, small servings of everyday nature, occasional deeper doses on a weekly or monthly basis, and rare, immersive experiences that occur seasonally or annually. All of these layers matter. Together, they create a *balanced nature diet*.

Here's one way to picture it:

1. Everyday Nature — The Base

These are microdoses of nature woven into daily life:

- a tree you pass on your walk to work
- a houseplant near your desk
- a window that looks onto sky or green
- five minutes of fresh air on the porch
- a short stroll through a pocket park

Even brief contact with nature can bring small but meaningful shifts: a softening of stress, a gentle restoration of attention, and a quiet sense of being connected to something larger than screens, schedules, and tasks.

Many of us walk outside while listening to a podcast or talking with a dear friend, and there is nothing wrong with that. But some of nature's deeper restorative gifts seem to arrive when we occasionally let our senses fully come forward. When we do, we begin to notice the movement of wind, the scent in the air, the texture of the ground beneath our feet, and the subtle rhythms of the natural world all around us.

2. Weekly Nature — A Little Deeper

Weekly doses might look like:

- a two-hour walk in a larger park or along a trail
- a morning at a botanical garden
- a Sunday picnic in a city greenbelt
- regular outdoor exercise—jogging, cycling, yoga in the park

What longer time outdoors often makes possible is a fuller exhale of the nervous system, the grounding rhythm of movement, and enough spaciousness for conversation, reflection, and play to unfold without being rushed.

3. Monthly Nature Escapes

These might be:

- day trips to a state park or quiet shoreline
- a half-day hike in a less-developed area
- a visit to a nearby forest, mountain, or wetland

There is a particular kind of opening that comes when we step outside the boundaries of our usual routines and neighborhoods. Larger landscapes, deeper quiet, and the presence of other lives around us can loosen the grip of what once felt enormous. What pressed in on us indoors may still be real, but it no longer fills the whole horizon.

4. Seasonal or Annual Wilderness Immersion

At the top of the pyramid live the big experiences:

- multi-day camping or backpacking trips
- multi-day nature retreats or pilgrimages
- extended time on a river, in the mountains, or by the sea

At the deeper layers, nature does more than restore attention. It can rearrange us. Values shift. Priorities clarify. What seemed urgent begins to soften, and what is truly important steps forward again. Just as quietly, a deeper sense of kinship may return—a felt knowing that we are part of a wider living world, not separate from it.

The point is not that everyone must reach the top each year. It is that every layer matters, and we do not need the summit for nature to begin its work.

Just as a healthy diet is not made of holidays alone, a nature-informed life isn't built only on rare retreats. It is built on rhythm.

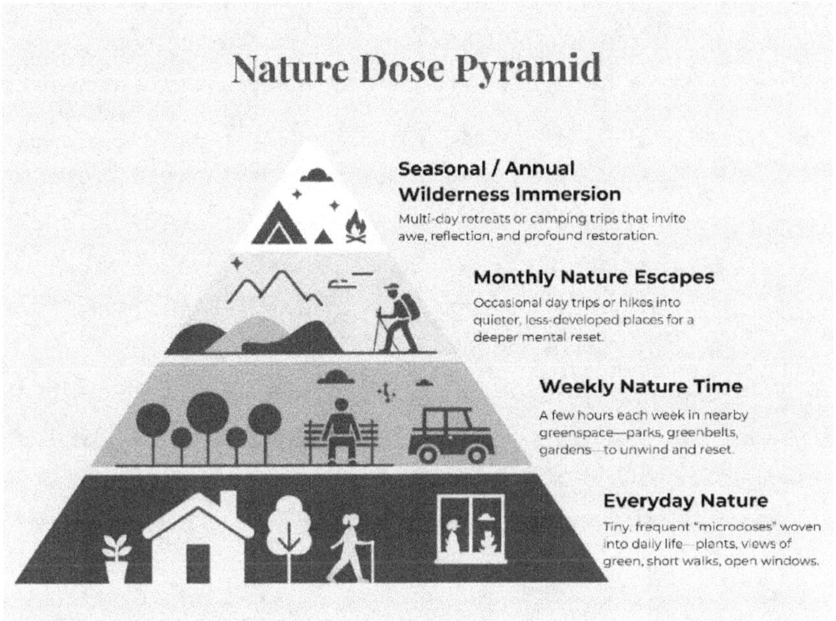

The Daily Microdose — A Day in Sarah's Life 🌿

Consider Sarah, a marketing manager whose calendar looks like many of ours: full days, long hours, constant digital contact. She loves the idea of nature but laughs when people suggest a week-long backpacking trip.

"I'm lucky if I get half a day off," she says.

Instead of chasing a perfect getaway, Sarah experiments with microdosing nature:

- Morning: She drinks her coffee by the window, watching birds at a feeder instead of scanning her phone. A five-minute ritual, but a different kind of start.

- Workday: She moves a few plants onto her desk and opens the window when the weather allows. On Zoom calls, she occasionally looks at the tree outside instead of her own image.

- Commute: She takes a slightly longer route home so she can walk through a small park, spending ten minutes under trees instead of on the sidewalk next to traffic.

- Lunch: When possible, she eats outside on a bench, even for ten minutes. No laptop. Just food, air, and whatever sky is visible.

- Evening: Before bed, she steps outside for a few breaths and watches the changing colors of the sky—sunset, twilight, or city lights against clouds.

None of these moments would look dramatic from the outside. They ask for no special gear, no retreat, no clearing of the calendar. But over time, Sarah begins to feel the difference. By five o'clock, she is less wrung out. When she loses momentum, creativity finds its way back more easily. And perhaps most importantly, she starts to recognize the difference between needing rest and needing renewed aliveness.

The stressors themselves have not disappeared. But nature has offered her nervous system something small and faithful: repeated chances to come back into balance.

This is the quiet wisdom of dosage—letting nature meet us where we actually are, rather than where we wish we were.

Practical Exercise – For All Practitioners

Creating Your Nature-informed Self-care Plan

Purpose

This practice helps you design a nature-based self-care plan that fits your real life, using the Nature Pyramid (and SHIFT Rx-style "challenge pyramid") as a loose guide.

You can use it for yourself, with clients, or in group or team settings.

Step 1: Take Stock of Your Current Nature Diet

Ask yourself (or invite others to journal):

- Daily: How often do I see or touch something living (tree, plant, sky, water) on a typical day?

- Weekly: Do I have any regular outdoor rituals—walks, exercise, family time?

- Monthly/Yearly: When was the last time I spent half a day or more in a wilder place?

Note what is already working. No judgment, just an honest snapshot.

Step 2: Name Your Self-Care Needs

Complete a few sentences:

- "Right now, my nervous system mostly feels…" (overwhelmed, flat, wired, dull, scattered, numb, etc.)

- "If I'm honest, I most need more…" (rest, excitement, connection, quiet, movement, perspective).

Then ask:

"How might nature support these needs?"

For example:

- Rest → shade, soft ground, gentle sounds

- Excitement → waves, wind, hills, exploration

- Connection → shared walks, picnics, gardening with others

- Perspective → big views, starry skies, long horizons

Step 3: Remember Times Nature Helped

Think of 1–3 memories:

- a walk that cleared your head

- a trip that changed how you felt about something

- a small moment—a sunset, rainstorm, or tree—that moved you

Ask:

- "Where was I?"

- "What was my body doing?"

- "What kind of nature dose was that—micro, day trip, or immersion?"

These memories are clues. They point toward what works for you.

Step 4: Choose Your Doses: Daily and Weekly

Start with one small daily practice. Specifically, vague intentions ("I'll go outside more") rarely survive a busy Tuesday. What works is something concrete, anchored to a habit you already have:

- "Every weekday morning, I will step outside for three minutes before touching my phone."

- "At lunch, I will sit near a window and look at the sky for ten slow breaths."

- "On workdays, I will walk the long way past the big maple tree instead of taking the shortcut."

I'll offer one of my own. At one point, I made a simple commitment: to drink my morning tea on the back porch instead of at my desk. It almost felt too small to count. And yet, after a couple of weeks, the porch had become its own kind of ritual. I began to notice the morning light differently. I started hearing the cardinals, who had likely been singing there all along. The daily microdose does not need to be dramatic. It only needs to be real enough for the body and senses to recognize.

From there, it can help to add a weekly or monthly rhythm— something slightly longer, with a little more depth. This might mean setting aside a Sunday afternoon for an hour or two in a park or on a trail. Or it might mean taking a half-day each month in a more spacious setting: a state park, a beach, a forest, a field.

What matters is that it becomes more than a nice idea. Put it on the calendar as you would any appointment that matters. Otherwise, it is surprisingly easy for even the things that nourish us to be pushed aside.

Step 5 – Name Your Barriers (and Be Kind)

Common barriers often sound like this: "I don't have time," "I don't have a car," "I don't feel safe in certain outdoor spaces," or "I forget until it's dark." For each, you can brainstorm a small adjustment, such as combining nature with something you already do like your commute, phone calls, or family time; exploring nearby parks, courtyards, rooftops, or community gardens if transportation is limited; choosing to go with a friend, seeking out populated and well-lit places, or using courtyards and balconies when safety is a concern; and setting a gentle reminder on your phone or linking the practice to an existing habit when forgetting tends to get in the way. Be gentle. The goal isn't to be perfect; it's to be a little more in relationship with the living world than you were last month.

Step 6: Write Your Plan as a Simple Promise

Keep it short. For example:

"For the next three months, I will:

- take one three-minute nature pause each morning on my porch;

- spend at least one hour each weekend in a park or green space;

- plan one day trip to a wilder place each season."

You can revise as you go. Plans are living things.

Chapter Highlights — What to Carry Forward 🌿

- We often fall into "someday retreat" thinking—believing only long wilderness trips will help—then feel stuck when we can't make them happen. In reality, small, consistent doses of nature are powerful.

- The Three-Day Effect shows that about three days unplugged in nature can significantly improve creativity, problem-solving, and mental clarity—like cleaning a mental windshield.

- The Wilderness Effect suggests that civilization's grip on our nervous systems is only a few days deep; extended immersion (four days or more) can awaken a sense of aliveness and reconnection.

- Nature immersion—backpacking, retreats, river trips, focused sit-spot time—can reduce anxiety and depression, deepen reflection, and restore a felt bond with the more-than-human world.

- The Nature Pyramid reminds us that we need a *range* of nature experiences: daily microdoses, weekly green time, periodic day trips, and occasional longer immersions. Each layer supports well-being in different ways.

- A "nature-informed self-care plan" doesn't have to be elaborate. One daily microdose, one weekly or monthly mini-reset, and one occasional immersion can profoundly shift how our nervous systems move through the year.

- Optimizing dosage is about working with the lives we actually live. When we stop waiting for perfect conditions and begin to weave nature into our everyday rhythms, we discover that healing doesn't just happen on distant mountaintops—it happens on porches, sidewalks, and in small parks, one breath of fresh air at a time.

In the ROOTED framework, Optimize Dosage invites us to think like gardeners: small, steady watering, with deeper soakings now and then. Nature is ready to meet us at any scale. Our task is simply to show up.

Chapter 9: T — Tend Relationships: Roots that Help Us Stay Standing 🌿

In this chapter, we turn toward the "T" in ROOTED: Tend Relationships.

We'll move into:

- why attachment is not a luxury but a survival strategy
- how loneliness erodes health and hope
- what it means to extend attachment theory beyond people to include place and Earth
- how nature can help repair attachment wounds, especially for those who were never securely held
- the unique power of outdoor groups, campfires, and shared landscapes
- how "place attachment" and grief for lost places show up in our work
- a practice for building a real, ongoing relationship with nature—like you would with a beloved friend

The Cedar and the Storm 🌿/💭

There is a story about a cedar tree that outlived the storms.

In a forest battered by wind and winter, one cedar stood tall year after year. Its trunk was scarred, bark split in places where lightning had kissed it. One day, a small bird landed on its branch and asked:

"How do you remain standing when the wind howls and the sky rages?"

The cedar answered:

Chapter 9: T — Tend Relationships: Roots that Help Us **Stay** Standing 🌿

"It isn't my strength alone.

Beneath the soil, my roots are braided with my neighbors'.

We hold the ground for one another.

If I stood here alone, I'd be gone.

It's our rootedness that keeps me upright."

We like to imagine ourselves as lone trees—independent, self-sufficient, unshakable. But most of us know, somewhere under the bark, that we are more like the cedar: we stand because something, or someone, holds us when the wind picks up. Ecologist Tom Wessels often reminds us that trees rarely grow in isolation. In healthy forests, their roots intertwine and fungal networks in the soil connect them in quiet exchanges of nutrients and information. Older, well-established trees can even help support younger or struggling neighbors through these underground partnerships. The forest, in other words, is not a collection of solitary beings but a community—one where resilience emerges through relationship.

Relationships—with people, with places, and with the world beyond the human-made—are not incidental. They are part of what steadies us. They form the root system beneath a life.

This chapter asks what happens when we begin to see connection not as backdrop, but as part of the healing itself. And deeper still, what happens when nature is invited into the story of attachment?

The Roots of Connection: Attachment 🌿/🩺

When my family discovered the TV show *Alone*, we were hooked.

Contestants are dropped into remote wilderness with a limited set of tools. They build shelters, catch fish, navigate predators and weather. Many are skilled survivalists. They know how to trap, how to start a fire in the rain, how to ration food.

136

Yet over and over, people tap out not because they can't find enough calories, but because they can't bear the aloneness.

They talk to the camera about missing their partners, their children, their community. Some cry as they watch the sun set over stunning landscapes. Others describe a kind of emptiness that feels more dangerous than hunger.

It's a reminder of something attachment theory has been saying for decades:

We are not designed to thrive in isolation.

Bowlby described attachment as one of our deepest built-in systems, the instinct that draws us toward a trusted other when distress rises. In childhood, this usually takes the form of reaching for a caregiver. In adult life, it may mean turning to a friend, a partner, a spiritual figure, or a beloved place that helps us feel steady.

When these bonds are secure enough, they become both refuge and launching point: a safe haven in difficult moments, and a secure base from which we can explore the world. When they are not, we find ways to adapt. Still, those adaptations often ask something of us.

Modern life strains attachment in ways that are easy to overlook. We move often. We live inside work cultures that prize independence, productivity, and constant responsiveness. We rely on technologies that can multiply contact while thinning its depth. And so it becomes possible to be surrounded on all sides, yet still carry a deep and private loneliness.

Loneliness: A Quiet Health Crisis 🌿

U.S. Surgeon General Vivek Murthy has called loneliness an epidemic, noting that chronic loneliness can shorten life expectancy as much as smoking fifteen cigarettes a day. That's not a poetic exaggeration; it's a public health comparison.

The BBC's 2018 Loneliness Experiment, which gathered responses from over 55,000 people worldwide, found:

Chapter 9: T — Tend Relationships: Roots that Help Us **Stay Standing** 🌿

- loneliness is common at many ages, not just in older adults

- people in cultures that prize independence (like the U.S. and parts of Northern Europe) are less likely to talk about feeling lonely

- the less people share their loneliness, the more isolated they feel

Meanwhile, the long-running Harvard Study of Adult Development—about 85 years and counting—keeps arriving at the same conclusion:

The strongest predictor of health and happiness in later life is not wealth or status, but the quality of close relationships.

Underneath all the data, the message is simple and ancient:

We are herd animals. Pack animals. Grove animals. Herds, packs, and groves get sick when they are cut apart.

Nature-Informed work takes this seriously. It recognizes that tending relationships—human and more-than-human—is not a side topic. It is central to resilience.

Expanding Attachment: When Nature Becomes a Secure Base 🌿/🩺

Attachment theory, as developed by Bowlby (1969), was originally about people, specifically, children and caregivers. But the core idea can be widened.

At its heart, attachment asks a few simple but profound questions: When I am distressed, where do I turn? And what serves as my secure base—the place, person, or presence that helps me feel regulated enough to venture out, explore, and engage with the world again? For many, nature has filled that role.

Think of the woman who returns to the ocean whenever life unravels. Standing at the shoreline, listening to waves, she finds perspective she cannot access anywhere else. The ocean doesn't speak, but it says plenty:

You are small and you belong at the same time.

This is bigger than your current crisis.

You have survived many tides before.

Or the person who always goes to the same patch of woods after hard therapy sessions. The trees become witnesses. The path remembers tears and breakthroughs. Over time, that short walk begins to feel like an embrace.

In these moments, nature is no longer just scenery. It begins to offer some of what we most long for in secure attachment: presence that does not disappear, acceptance that does not tighten or recoil, and a kind of soothing that reaches us through the senses before words are even possible.

A river remains, whether the day has gone beautifully or badly. A tree does not turn away because grief, anger, or joy found expression beneath its branches. The sound of water, the scent of earth, the feel of bark or wind or sunlight on the skin—all of these can help the nervous system remember how to settle.

We might think of this as attachment to place, and perhaps even attachment to Earth.

Then nature becomes part of the answer to one of the oldest questions we carry: *Where do I go when I need to feel held?*

Attachment Repair in Nature 🌵/🌿

For many clients and participants, early attachment stories were not gentle. They may have learned: You are lovable only when you behave perfectly. Your needs are too much. You will be abandoned if you show weakness.

Even as adults, those messages echo. Many carry a quiet conviction: If people really knew me, they would leave.

With that history, turning toward others can feel risky. Vulnerability is laced with danger. Therapy that relies solely on human relationship may feel both necessary and terrifying.

Nature offers a different kind of attachment experience. When someone steps into a forest, sits beneath a tree, or lies in the grass, they encounter presence without evaluation, company without commentary, a kind of unconditional being-with.

The tree does not require cheerfulness. The river does not withdraw if someone is angry. The mountains do not say, "Call me when you're in a better mood." For some people, this is the first relationship in which they are not asked to perform.

Clients sometimes say it plainly: "The woods don't care if I'm a mess." "I feel like the river can handle my grief." "When I sit with that tree, I don't feel judged."

Of course, trees are not therapists. But nature can still support attachment repair—by modeling nonjudgmental presence, offering consistent sensory cues of safety, and giving people a place to practice being fully themselves in the company of another living presence.

As therapists and helping professionals, we can name this explicitly:

"It makes sense that people feel risky. Notice how this place responds when you bring your real feelings here. What do you feel from the land itself?"

Nature becomes a co-regulator and a co-witness.

An Interdependent Bond: Indigenous Knowing 🌿

For many Indigenous cultures, this way of relating to land is not a new idea. It is the starting point.

Land is not scenery; it is kin. Self and place are braided together. Identity includes mountains, rivers, animals, winds.

Aboriginal perspectives, for example, often emphasize an ecological self: being a person means being part of Country, of a specific land, with specific ancestors, stories, and responsibilities. From early childhood, people learn to see land as a source of guidance and comfort, to understand certain places as sacred or protective, and to feel accountable to particular ecosystems and beings.

This is more than "loving nature." It is an interdependent bond, where well-being flows both ways: people care for land, and land cares for people.

As non-Indigenous practitioners, we do not appropriate these worldviews. But we can acknowledge that Western separation of "self vs. environment" is not universal; recognize that many clients carry cultural or ancestral understandings of land that differ from our own; and remain open to learning from Indigenous scholars, elders, and communities who are willing to teach.

When we view attachment through this wider lens, tending relationships means tending human bonds, place bonds, and ancestral and cultural bonds. Healing may come not only from "working on our issues," but from remembering ourselves as part of Earth's larger family.

Gathering Around the Fire 🌿

If attachment to place is one root, attachment to community in place is another.

For most of human history, people gathered outdoors—around water, in fields, beneath trees, and especially around fire. Heying and Weinstein call the campfire "a forge for ideas." It is also a forge for relationships. Around flames, people share stories, pass down wisdom, laugh, argue, cry, sing, and sometimes sit in silence without needing to fill it. A fire can function almost like a natural meditation object—something simple and ancient that draws the mind into the present moment.

Chapter 9: T — Tend Relationships: Roots that Help Us **Stay** Standing ⚘

The glow on faces, the shared warmth, the simple act of seeing one another in a circle: this is attachment work in its oldest form.

Modern life has its own circles: meetings, video calls, classrooms. But something different happens when the circle is outside. Hierarchy softens; everyone is equally lit by the same fire or sun. Nervous systems settle with the rhythmic crackle of wood or the chorus of night sounds. Conversation can move more easily between light and heavy, play and depth.

When groups gather in nature—even without a literal fire—they often rediscover something their bodies already know:

We are meant to face each other, not only screens. We are meant to share the same air.

Why Outdoor Groups Feel Different ⚘/✿

As a therapist who has led both indoor and outdoor groups, I notice a consistent pattern.

Indoor groups are valuable. People do deep work in rooms. But trust often builds slowly. People can feel watched or self-conscious in tight spaces, and the room itself can echo other institutional settings—school, hospital, office.

Outdoor groups move differently. Walking side by side on a trail, people often speak more freely than they do face to face in chairs. Sitting under trees, participants may pause longer after someone shares something vulnerable—as if the group is letting the wind, not the clock, decide when to move on.

I've seen reserved participants share more on a single hike than in weeks of office sessions. I've watched conflicts soften when addressed while moving together rather than sitting still. And laughter comes more easily when the group gathers around a picnic table—or a fire.

Why? Nature seems to reduce performance pressure ("it's not all about me; there is a whole landscape here"). It offers external anchors—

birds, sky, water—so people can look away and regulate without losing connection. And it creates a shared experience, terrain, sound—that bonds people without forcing intimacy.

For group-based NIT and NIC, outdoor settings can accelerate trust, deepen empathy, and help people feel part of something larger than their own pain. We still hold safety, confidentiality, and clinical boundaries. Nature simply joins the facilitation team.

Nature Claps for Courage 🌿/💬

During a nature-immersion retreat in the wooded hills of Western Maryland, our group gathered in a quiet circle beneath a canopy of oaks and black walnut trees. We were practicing the Way of Council: passing a talking piece so each person could speak from the heart while the rest of us listened with our full attention.

When the talking piece reached a young woman in her mid-twenties, she held it as if it were fragile. Her fingers trembled. She took a long, steadying breath and said, almost apologetically, that speaking in groups had always terrified her.

She paused, eyes fixed on the ground.

"But something about being out here," she added, glancing up at the trees around us, "makes it feel just safe enough to try."

Her voice shook as she shared a story she had never told aloud in front of others. The circle stayed quiet, not rushing to reassure or fix. We simply listened, letting her words move through the space between us.

When she finished, she placed the talking piece back in the center of our circle.

For a moment, there was that special kind of silence that follows honest speech: thick, tender, almost humming. No one quite knew what to say next.

And then the forest responded.

Chapter 9: T — Tend Relationships: Roots that Help Us **Stay Standing** 🌿

Behind us, a sudden burst of sound broke the stillness. We all jumped, then turned to see a cluster of black walnuts dropping from the branches above the parking lot. They hit the pavement one after another with sharp cracks.

It sounded, unmistakably, like applause.

The whole circle broke into smiles. Someone laughed. The young woman's eyes filled with tears.

"I guess nature approves," someone whispered.

Of course, we could have explained it away. Walnuts fall in autumn. Trees do what trees do. And yet the timing entered the body before the mind could dismiss it. For that young woman, the sound arrived as a kind of affirmation—not from a therapist offering a well-timed reflection, and not from a group hurrying to reassure her, but from the land itself.

Moments like this are not techniques we can produce on command. They are gifts. What we can do is tend the conditions that make such moments more likely to be noticed. We can create circles of deep listening. We can slow down enough to feel the living world around us. We can allow nature to enter not only as backdrop, but as witness and companion.

And when we do, people sometimes receive exactly the encouragement they need—from a falling walnut, a shift in light, a sudden gust of wind, or a bird that chooses that moment to sing.

Nature has its own way of clapping for courage.

Attachment to Place

Ecologist Tom Wessels describes a remarkable phenomenon found in some forests: tree stumps that remain alive long after the trunk has been cut. Although the stump no longer has leaves and cannot photosynthesize on its own, neighboring trees keep it alive through interconnected root systems, sharing water, sugars, and nutrients underground. The living trees continue to sustain the stump, sometimes for decades. Wessels uses this example to illustrate a deeper truth about forests—that trees are not merely individual organisms competing for survival, but members of a cooperative community where life is often sustained through connection and attachment to place.

Place can function as an object of attachment, embodying qualities similar to those of attachment figures. These emotional qualities, such as a sense of safety, familiarity, and refuge, make places feel like a secure base—offering calm, restorative, and grounding experiences.

Key aspects of place attachment include:

- Consequences of childhood memories: The emotional connection formed in childhood can influence the places we feel drawn to.

- Places of refuge: A place that feels like a retreat, offering solace and calm.

- Calming, restorative, or felt sense of safety: Being in certain spaces that naturally calm and restore us, giving a sense of emotional regulation.

- Positive first-hand experiences: Experiences in a place that foster positive emotions and memories.

- Shared cultural ideologies of groups and shared interactions with place: Group affiliations or communal connections to specific places create a shared sense of belonging.

- Favorite recreational activities: Certain places we frequent for leisure activities, reinforce the bond with that space.

- Familiarity & nostalgia: The emotional draw of familiar spaces that invoke a sense of comfort or nostalgia, reminding us of a previous sense of safety.

- Sites of loss and tragedy: Some places may serve as a connection to past loss or tragedy, offering a unique kind of attachment that fosters reflection, healing, or remembrance.

Memories, time, and thoughts experienced while in a park or green space can encourage attachment to that place. People become attached to peaceful, restorative green spaces that offer mental and physical respite, incorporating them into their routines and even their self-identity. These spaces often fulfill crucial health needs, both mental and physical, and provide a sense of stability and renewal. Research shows that people feel more at ease in landscapes resembling those of their childhood, and that they experience measurable stress reduction when recreating in settings where they feel most at home. This connection demonstrates the importance of green spaces as restorative environments that nurture both body and mind (Wolf, K.L., S. Krueger, and K. Flora. 2014).

Place Attachment — When a Place Feels Like Home 🌿/🩺

For some people, our most reliable attachment figure is not a person but a place.

Place attachment research describes how certain environments come to feel like refuge, a place we go to rest, hide, and recover, and a secure base, somewhere we can return to again and again. Qualities that often build place attachment include:

- powerful childhood memories (grandma's garden, a lake, a neighborhood park)
- places of refuge during hard seasons (a library, a particular bench, a forest path)
- landscapes that resemble "home" in early life (people often feel calmer in places similar to the environment they grew up in)
- shared spaces where cultural or community rituals happen
- green spaces where people regularly move, breathe, and think

Over time, places can become part of who we are. They enter the language people use to describe themselves: someone formed by mountains, someone who finds the ocean church-like, someone who knows exactly which park to return to when life begins to unravel.

Physiologically, green spaces are linked with lower stress, more movement, and improvements in mood and concentration. But the emotional offering may be just as profound. A place can tell us: there is somewhere you can go. There is somewhere you belong. You are not without roots.

As practitioners, we can approach this gently, with questions that help people remember where this kind of attachment already exists. Is there a place, indoors or outdoors, that has felt like home? Where does your nervous system breathe a little more freely? Often, the answers reveal that attachment to place is already present, waiting for language.

When Places Are Lost — Grief for Land and Home 🪴/🌵

Attachment to place also means we grieve when place is harmed or taken. People may experience intense loss when a childhood home is sold or demolished, when a beloved forest is clear-cut or burned, when a favorite park is paved over for development, or when war, climate disasters, or economic forces uproot them from their land.

This grief can look like nostalgia that cuts deeper than "missing a view," anger or despair at environmental destruction, a sense of being unmoored in a new city or country, or difficulty forming attachments to new places.

During a training with staff at an asylum and refugee organization, we were focusing on grief-specific approaches. When we came to place attachment, something clicked. One person said, "We've been attending to grief for people, but not grief for land."

Many of their Nigerian clients were mourning not only lost family and stability, but the sensory and cultural textures of home: familiar smells, foods, and weather; the colors of soil and vegetation; the sounds of birds, languages, and markets; spiritual and cultural relationships with particular landscapes. Washington, D.C., felt nothing like home.

We spent time brainstorming ways to honor grief for the old land while also supporting new place attachments in the current city: parks, rivers, community gardens, cultural spaces.

One participant said, beautifully: "Mother Earth wears many faces across the world, yet she remains the same nurturing presence. Being in D.C. is simply learning to recognize another of her expressions."

Helping people reconnect with Earth where they are now doesn't erase what was lost. But it can offer a path toward groundedness and rootedness in unfamiliar soil.

Practice: For All Practitioners

Translating Human Connection into a Relationship with Nature

Purpose

To help participants build an intentional, caring relationship with the natural world using the familiar steps of forming a relationship with a person.

You can use this with individuals, groups, or for your own reflection.

Step 1: "Meeting" Nature with Fresh Eyes

Choose one specific place:

- a city park
- a trail
- a single tree
- a courtyard garden
- even a balcony with plants

Arrive as if you were meeting a new person. For 5–10 minutes, simply be there.

Ask yourself:

- What do I notice first—shapes, colors, movement?
- What does this place "say" through its sounds or stillness?
- How do I feel in its presence—tense, calm, bored, curious?

Reflection:

When we meet someone new, we give them room to be themselves before deciding who they are. How might you offer that same patience and openness to this place?

Step 2: Learning Nature's Story

Choose one element: a tree, rock, patch of grass, body of water.

Wonder about it:

- How long has this been here?
- What has it "seen"?
- How has it changed over time?

Later, if you wish, look up:

- what species it is
- how long such trees or stones typically endure
- the history of the park or area

Reflection:

How does learning someone's background deepen your connection to them?

What shifts when you know the story of this place?

Step 3: Building Respect and Trust

Healthy relationships rest on mutual respect.

Sit quietly for 5 minutes in this place.

- Notice what you hear, see, and feel.
- Resist the urge to pick, pluck, or take anything.
- Practice being with, not doing to.

Ask:

- How might my presence here affect this place?

- What does respect look like in this relationship?

Reflection:

When someone respects your limits and needs, how does that feel?

How might you show similar respect to this place?

Step 4: Touching with Care

If it feels right and safe, explore a small physical connection.

- Touch the bark of a tree.

- Run your fingers over moss or stone.

- Hold a leaf or pinecone gently.

Before touching, silently ask permission in your mind:

"Is it okay if I connect with you this way?"

Listen for your body's response.

Reflection:

How does respectful touch—human or more-than-human—change your sense of belonging?

What do you feel in your body when you connect with this place through touch?

Step 5 – Celebrating Milestones Together

Relationships deepen when we mark time and milestones.

Notice the "milestones" in your chosen place:

- first buds or leaves in spring

- the height of summer green

- the colors of autumn

- the quiet changes of winter

- a first snowfall, heavy rain, or strong wind

Choose one event and celebrate it:

- pause and really watch it

- write a short poem or note about it

- bring a friend to celebrate the solstice

- offer a small, non-invasive gesture (a song, a moment of gratitude)

Reflection:

How does celebrating nature's changes shift your sense of connection?

Does this place feel more like a "someone" than a "something" over time?

Step 6: Building a Shared Future

Relationships need ongoing care.

Ask:

- How do I want to continue relating to this place?

- What small, concrete actions can I take to nurture this bond?

Examples:

- visiting regularly

- picking up litter when you see it

- supporting local conservation efforts

- bringing others here to meet this place

Write a short promise or intention:

- "I intend to visit you at least once a month."

- "I will help keep you clean."

- "I will pay attention to your changes."

Let it be simple and sincere.

Chapter Highlights — What to Carry Forward 🌿

- Humans are wired for connection. Attachment theory reminds us that secure relationships are not optional extras; they are central to resilience and well-being.

- Loneliness is a serious health risk, associated with shortened lifespan and increased physical and mental illness. Our work, especially in NIT and NIC, is partly about weaving people back into webs of connection.

- Attachment can extend beyond people. Many find secure base qualities in nature—specific trees, rivers, coastlines, or parks that offer comfort, perspective, and a sense of being held.

- For those with painful early attachment histories, nature can provide a nonjudgmental presence, helping to repair internalized messages of unworthiness and conditional love.

- Indigenous perspectives remind us that land and self are deeply intertwined. We honor, not appropriate, these worldviews by recognizing that identity and place are often inseparable.

- Outdoor groups and gatherings around fire tap into ancient patterns of social bonding. Nature settings tend to accelerate trust, deepen conversation, and soften defenses.

- Place attachment explains why certain places feel like home—and why we grieve when they are lost or damaged. Naming grief for land and home (especially with refugees, displaced people, and those facing environmental loss) is part of good care.

- Practices that treat nature like a partner—not a tool—help people develop healthier, more reciprocal relationships with the Earth. Tending these bonds can become a powerful source of strength for both clients and practitioners.

In the ROOTED framework, Tend Relationships means tending the whole web: self, others, place, and more-than-human kin. Our job is to

help people remember they were never meant to weather the storm
alone.

Chapter 9: T — Tend Relationships: Roots that Help Us **Stay** Standing

Chapter 10: E — Enact Reciprocity & Meaning: Living as If the Earth Loves You Back 🌿

In this chapter, we arrive at the "E" in ROOTED: Enact Reciprocity & Meaning.

So far, we've focused on regulation, attention, senses, dosage, and relationships. Here, we go one step further. We ask:

If nature is not just medicine but relationship, how do we live like that is true?

This chapter invites you to explore:

- awe and wonder as quiet antidotes to despair

- the shift from ego-centered to eco-centered awareness

- spiritual transcendence and what happens when the self feels bigger and smaller at the same time

- eco-spirituality: seeing Earth as sacred partner, not scenery

- the Honorable Harvest as a guide for ethical, reciprocal living

- simple rituals—like tea and herbs—that bring reciprocity into therapy and daily life

- a *Cosmic Perspective* practice to root responsibility in awe, not guilt

The Beaver Who Blessed Our Circle 🌿/💭

It was day four of a backpacking trip in Dolly Sods Wilderness, a high plateau of rock, bog, and sky in West Virginia. The group was made up of therapists—people used to hold space for others, used to talking about feelings indoors under fluorescent lights.

155

Chapter 10: E — Enact Reciprocity & Meaning: Living as If the Earth Loves You Back 🌿

We had left cell signal behind days ago. Blisters and tired shoulders had softened people's defenses. Meals cooked over tiny stoves tasted better than any restaurant.

That evening, we gathered near a cluster of small ponds for a sit-spot practice. Each person found a place to sit in silence. The sky slowly slid toward gold. The water turned into a mirror.

About five minutes in, I saw movement.

A large beaver slipped into the water and swam toward us. It did a slow, unhurried lap near the group, then another. Then, with a sudden smack, it slapped its tail on the surface—loud enough to startle us—before gliding away.

I looked around.

Several of the therapists had tears on their cheeks. Later, in our evening circle, words came slowly. What they reached for was not explanation so much as recognition: that it had felt like a blessing, as though the land had acknowledged our presence; that for one person, who had long felt like an outsider in nature, something had shifted and inclusion had become briefly, unmistakably real.

The beaver had not come *for* us. And yet it arrived within our experience in a way that altered it. For a brief moment, the boundary between "human" and "nature out there" loosened. The group felt not just surrounded by nature, but claimed by it.

That is one of awe's quiet powers. It is never only *wow, that was beautiful.* It changes the inner arrangement of things.

Awe – When the World Is Bigger than Our Worries 🌿/🐍

Psychologist Dacher Keltner describes awe as:

"The feeling of being in the presence of something vast that transcends your understanding of the world."

Awe can rise up in response to nature—storms, mountains, starry skies, even a beaver doing laps at sunset. It can also come through art and music, moral beauty (witnessing courage, kindness, or sacrifice), or collective movement: a march, a choir, a ceremony.

What makes awe special is not just that it feels good. It re-sizes us.

Biologically, awe appears to quiet activity in the brain's default mode network, the circuitry tied to self-focused rumination. It softens the borders of "me, my problems, my story," and nudges us toward humility, kindness, and generosity. Keltner's research also suggests awe can shift us from "I" to "we," increase prosocial behavior, and deepen our sense of purpose and meaning.

In the Dolly Sods beaver moment, no one was thinking, *What's my five-year plan?* They were thinking: *I'm part of something intricate and alive. This place is aware of me, too.*

For clients who feel stuck inside their own distress, awe can crack open a window—not by denying pain, but by reminding them that pain is happening within a much larger, mysterious world.

From Ego to Eco – Frank's Awe Walks

Frank arrived in therapy knotted with worry.

He worried about work, reputation, finances, health, and whether he was somehow doing life wrong. His attention had narrowed into something like a flashlight fixed almost entirely on his own mind, illuminating every thought, every fear, every possible misstep.

So we began a simple practice: a daily Awe Walk.

The instructions were spare. Once a day, he would take a short walk—around the block, through a park, even down an ordinary street. He was not asked to achieve anything or to feel a certain way. His only task was to notice one moment that stirred awe or wonder. It could be something very small: light catching on a leaf, a child's laughter, the shape of clouds, a pattern in cracked concrete.

Chapter 10: E — Enact Reciprocity & Meaning: Living as If the Earth Loves You Back

On his first Awe Walk, Frank noticed a butterfly hovering over a yellow flower. He could not have explained why, but he stopped.

His breath slowed. For a few seconds, worry loosened its grip. He was no longer thinking about email, income, or whether he was somehow behind in life. He was simply watching a small, delicate creature move through the world as if it fully belonged there.

The next day, he noticed an elderly couple feeding birds by a pond, the ease of their routine speaking between them. Another day, a red-tailed hawk passed overhead and vanished into the sky. Gradually, these moments began to gather.

Later, Frank told me, "It's as if my life had become a room where the only window faced my anxieties, and Awe Walks cut a second window into the wall."

He did not become carefree. But he became less tightly organized around himself and more aware of the world as a shared and living field.

This is one of awe's quiet gifts. Practiced over time, it can shift us from an ego-centered way of seeing to one that is more ecological, more relational, and more alive.

Spiritual Transcendence – When the "I" Expands and Thins

Awe often opens the door. Spiritual transcendence is what can unfold when we step through it.

If awe is the moment of wonder, transcendence is the deepening that sometimes follows—a more enduring shift in which we begin to feel ourselves as part of something larger, something we did not create and cannot fully contain. Our lives start to feel threaded into a wider story. Neuroscientist Andrew Newberg's brain-imaging research suggests that during transcendent states, the regions involved in maintaining sharp boundaries between self and world may quiet. People often describe a felt sense of unity—with nature, with the universe, with the divine. They speak of a love or wholeness that resists easy language, and of a reordering of priorities: less fixation on status, more care for others, for the Earth, and for what truly matters.

In my own research, I found a similar pattern. People who feel more connected to nature often report better mental health. Nature connectedness is also frequently linked with spirituality. And spiritual transcendence—a felt relationship with something greater—seems to offer something more still, adding benefits beyond nature connectedness alone.

In other words, it is powerful to feel connected to trees, rivers, and animals. It may be even more powerful when that connection is experienced as part of a larger sacred fabric.

This is not about adherence to a particular religion. It is about a movement in perception. We begin by seeing ourselves as standing over nature, using it. Then perhaps we come to feel ourselves within nature, supported by it. And sometimes, deeper still, we begin to sense that we belong to a reality spacious enough to include both nature and something beyond what we can fully name.

Ram Dass once said, *"We're all just walking each other home."* Spiritual transcendence is what happens when our sense of home begins to widen, from a physical house, to a shared Earth, to something larger still.

Eco-Spirituality – Nature as Sacred Partners

For centuries, many cultures have treated nature not as backdrop but as sanctuary: forests as cathedrals, mountains as places of revelation, rivers as the veins of the Earth.

Eco-spirituality builds on this understanding: all life is sacred. The Earth is not a thing we live *on*, but a being we live *with*.

This perspective invites a different kind of attention. We begin to notice where reverence already lives in our encounters with the natural world. We may find ourselves wondering whether there is a place that feels holy to us, even if we would never use that word aloud. We may begin to see how grief, healing, and meaning-making are already woven into our relationship with land.

The wider world teaches spiritual truths in ways that are both ordinary and profound. Trees show us how to be rooted without becoming rigid, supported by the ground and still able to move with the wind. The seasons remind us that loss and renewal are not opposites, but companions. Soil teaches its own quiet wisdom: that what falls away does not simply disappear, but transforms, returns, and becomes nourishment for what comes next.

When we begin to receive these lessons, eco-spirituality becomes less an abstract philosophy and more a way of being. We approach the Earth with humility instead of entitlement. We learn to hold our suffering within larger cycles. And slowly, the central question begins to change—from *What can I get from this world?* to *What is my role within this living community?*

One simple practice for this shift is the daily awe walk. Its purpose is not dramatic insight, but a gentle reorientation from ego to eco—from the self as solitary center to the self as participant in a wider web of life. By noticing one small moment of awe each day—a change in light, a bird call, an opening in the clouds, an unexpected act of kindness—we loosen the sense that we are sealed inside our own private story.

Over time, this simple practice can cultivate presence, gratitude, and a steadier felt sense of belonging.

The practice is simple: notice something that evokes awe, however small. Let it reach you. Breathe. Then pay attention to what shifts, even slightly, when your awareness touches something larger than yourself.

Does the Earth Love You Back? ✿/💬

Most of us are familiar with the idea of loving the Earth.

Robin Wall Kimmerer, the Potawatomi botanist, takes this understanding a step further. She writes:

"Knowing that you love the Earth changes you…

but when you feel that the Earth loves you in return,

that feeling transforms the relationship from a one-way street into a sacred bond."

It is worth pausing with that for a moment.

It is one thing to love a forest. It is something deeper, and perhaps more vulnerable, to feel loved by it in return—to feel steadied by its trunks, comforted by its shade, or quietly seen by its presence.

For many people, this shift is not abstract. They can point to particular moments: a time when the ocean seemed to meet them in grief, when a certain rock outcropping felt strangely holding during a dark season, or when a bird or animal appeared at just the right moment, carrying an uncanny sense of companionship.

We might be tempted to dismiss such experiences as magical thinking. But perhaps they speak to a different grammar of relationship—one that moves beyond *I appreciate nature* toward *I am in relationship with a living world that meets me in ways I may never fully understand.*

Khalil Gibran captures something of this intimacy when he writes that the earth delights in our bare feet and the winds long to play with our hair. When people begin to feel loved by the Earth in this way, something in them often softens.

Presence as Prayer – The Backyard Sit Spot 🌿/💭

My first sit spot wasn't in a national park. It was on a log in my backyard. I set a timer for ten minutes and sat. At first, I was bored. My mind made lists: emails to answer, chores, deadlines. I wondered if I was doing it "right." Then my attention snagged on a small movement: an ant carrying a piece of something far larger than itself.

I watched.

The ant struggled, stopped, adjusted its grip, then continued. It paused, rested, then tried again. After a while, another ant appeared and joined the work. A tiny, ordinary drama, but it began to speak. The ant showed me perseverance—moving one small burden at a time. It showed me rest—stopping when needed rather than pushing past capacity. It showed me community—help arriving, unannounced, when the load was too big.

At some point I realized the timer had long since gone off. Nearly thirty minutes had passed, and I hadn't noticed. I had been so absorbed in the moment that time seemed to soften and stretch. It felt like one of the most nourishing, restorative moments I had experienced in a long while.

Nothing about my outer life changed in those minutes. But my inner stance did. I realized I often expected myself to carry far more than felt possible, without rest. I forgot that help could appear—that I didn't have to do everything alone. And I overlooked how much wisdom lives in tiny, more-than-human lives.

This is the spirituality of presence: not dramatic visions, not complicated rituals, just steady attention to what is alive in front of us.

Mindfulness in nature is not about zoning out. It's about tuning in: to ants, wind, light, and the stories we learn when we are willing to be still.

The Honorable Harvest – Learning to Ask, Receive, and Give Back 🌿/♑

Robin Wall Kimmerer describes the Honorable Harvest as an Indigenous ethic of taking from the earth with respect and reciprocity. While specific teachings vary by Nation, common threads include:

- never taking the first of a plant you meet

- taking only what you need

- asking permission before harvesting

- expressing thanks

- using everything you take

- giving something back (care, offerings, restoration)

The Honorable Harvest is not merely a metaphor. It is a way of moving through the world—a daily ethic, a lived code of relationship.

Seen through a nature-informed and eco-spiritual lens, its wisdom becomes deeply practical. Before taking or using, we are asked to pause and consider: Is this truly needed? Is this the right source? We are invited to take less, resisting the habits of excess and leaving enough for others, both human and more-than-human. Gratitude becomes part of the practice too, whether spoken aloud or contained within. We are reminded to use well what has been given, rather than wasting it. And always, there is the call to return—to give back through restoration, conservation, activism, stewardship, or small acts of everyday care.

This ethic reshapes the meaning of Nature-Informed Care. We are not simply helping people feel better by using nature as a tool. We are inviting them into a reciprocal relationship, one in which healing is not extracted in a single direction, but allowed to move both ways.

My take: this keeps the structure of the original ideas, but now it reads like a reflective book passage rather than a principles handout.

Tea and Herbs as a Small Honorable Harvest

One simple way to weave the Honorable Harvest into counseling and group work is through herbs and tea.

Chapter 10: E — Enact Reciprocity & Meaning: Living as **If the Earth** Loves You Back 🌿

Imagine a small shelf, a mini apothecary, in your therapy room: jars of chamomile, peppermint, lemon balm, lavender; perhaps rosemary, tulsi, or ginger; a kettle, simple cups, and a quiet, deliberate ritual.

Before or during a session, you might say:

"Today, we'll start by choosing a plant ally for our time together. Each herb offers something different: calm, focus, grounding. See which one draws you."

Then invite clients to:

1. Choose an herb (or blend) that matches what they need that day:

    o   calm/sleep → chamomile, lavender

    o   clarity/energy → peppermint, ginger

    o   grounding/steadiness → rosemary, tulsi

2. Smell it first. Inhale; notice memories and sensations.

3. Offer thanks, quietly, before steeping.

4. Sip slowly at the start or end of the session, noticing warmth, flavor, and any shifts in the body.

As you do this, you can name the Honorable Harvest in simple ways: acknowledge where the herbs came from, talk about sustainable sourcing when appropriate, and invite clients to consider how they might "give back" (growing herbs, supporting local growers, reducing waste).

The tea becomes more than a beverage. It becomes a tangible experience of receiving from the Earth, a small ritual of gratitude and reciprocity, and a sensory anchor for calm, presence, and return.

For many, this kind of gentle ceremony can feel almost sacramental, even without the use of religious language.

For those who don't drink tea, the ritual can take other forms: cooking with herbs at home, adding mint or lemon balm to water, placing lavender by the bedside, or simply holding and inhaling a plant during moments of stress. The invitation remains the same: receive consciously, give thanks, and remember relationship.

Practice: For All Practitioners 🌿/💭

Cosmic Perspective: Expanding Our Awareness of the Universe

Purpose

To use the vastness of the cosmos not to make us feel meaningless, but to deepen humility, belonging, and responsibility to this small, luminous planet.

You can offer this as an individual exercise, a group guided practice, or a journaling prompt.

Step 1: Look Up

Find a place where you can see the sky:

- night sky full of stars (ideal)
- a clear daytime sky
- even a cloudy, shifting canvas

Stand or sit comfortably.

Let your eyes soften and rest on the sky for a minute or two.

Ask yourself:

- What does my body feel like when I remember that above this sky are stars, galaxies, distances I cannot truly imagine?

Step 2: Feel the Tiny, Precious Dot

Bring to mind images you may have seen of Earth from space:

- the "blue marble"
- the pale blue dot floating in darkness

Remember astronaut accounts of the Overview Effect—how seeing Earth from orbit rearranges their sense of identity and urgency:

- everything we love is on that fragile sphere
- borders vanish; only one planet is visible

Reflect silently or journal:

- *What happens in me when I remember that my entire life—everyone I love, every joy and sorrow—has unfolded on this one small, floating home?*

- *Does this make my concerns feel smaller, more precious, or both?*

Step 3: From Human-Centered to Life-Centered

Now widen your awareness:

- think of forests, oceans, deserts, coral reefs

- think of animals, insects, fungi, microscopic life

- think of the atmosphere, climate patterns, cycles of water and carbon

Let this sentence land:

"I am not at the center of this story. I am part of its cast."

Sit with that for a few breaths.

Ask:

- *If I am not the main character, who else's well-being matters in every decision I make?*

- *What kind of supporting character do I want to be in Earth's story?*

Step 4: Horton and the Speck

Think of Dr. Seuss's *Horton Hears a Who!* The elephant who protects an entire tiny world living on a speck of dust.

Imagine:

- Earth as that speck in the vast cosmos

- you as Horton, holding it carefully

Visualize:

- your hands cupping this small, shimmering planet

- every ecosystem, every culture, every child and elder resting there

Ask:

- *If the well-being of this speck mattered to me the way the Whos mattered to Horton, what is one small thing I would do differently today?*

Let an action arise naturally:

- picking up trash on your walk

- reducing one kind of waste

- voting or advocating for policies that protect water, land, or air

- caring for a local tree, garden, or habitat

- reaching out to someone in your community, because human connection is part of Earth's health too

Write it down as a simple promise.

Step 5: Bring the Cosmos Back into Your Day

To close, take three slow breaths:

1. Inhale – "I am small."

2. Exhale – "I am part of something vast."

3. Inhale – "My actions matter."

4. Exhale – "I choose to care for this speck."

Carry this awareness into your day—not as pressure, but as an gesture. In the days after a recent election result left me feeling discouraged, I found myself watching a conversation between Neil deGrasse Tyson and Stephen Colbert in which Tyson described what he calls the cosmic perspective. This view does not erase our struggles, but it reminds us that we are tiny participants in an unimaginably large story—one small species on a small planet orbiting an ordinary star in a vast galaxy. Holding that perspective for a moment helped me breathe a little easier, remembering that while human events matter deeply, they unfold within a universe far larger and older than any single political moment.

Chapter Highlights — What to Carry Forward 🌿

- Awe is more than a pleasant feeling; it quiets self-focused rumination and helps us feel part of something larger, which increases kindness, purpose, and resilience.

- Spiritual transcendence builds on awe, shifting us from a narrow "me" story into a felt experience of being part of something greater—whether experienced as God, Spirit, Universe, or simply the living web of Earth.

- Eco-spirituality treats nature as sacred partner, not backdrop or resource. It invites us to see forests, rivers, seasons, and creatures as teachers in how to live, grieve, and grow.

- Feeling that the Earth loves us back—as Robin Wall Kimmerer describes—turns a one-way appreciation into a mutual, healing relationship.

- The spirituality of presence shows up in small moments: a sit spot, an ant carrying its burden, wind in trees. Nature teaches perseverance, rest, and interdependence simply by being itself.

- The Honorable Harvest offers an ethic of reciprocity: ask, take only what you need, give thanks, use what you take, and give back. This can shape how we shop, eat, harvest, and design our programs.

- Simple rituals—like herbal tea chosen with intention—can weave reciprocity and gratitude into therapy and caregiving, making Nature-Informed work more relational and less extractive.

- A cosmic perspective doesn't erase our problems; it places them inside a larger story, often reducing despair and increasing a sense of responsibility and reverence for our tiny, beloved planet.

In ROOTED, Enact Reciprocity & Meaning is the final movement: letting nature not only heal us, but change how we live. When we act as if the Earth is part of our "us," every small choice—a walk, a cup of tea, a piece of trash picked up—becomes part of a quiet, ongoing prayer:

Thank you. I remember. I am trying to be good company for this world.

And in the next chapter, we'll turn to the D—bringing all of this into deeper, lived integration.

Chapter 11: D — Design the Setting: Let the Room Remember the Forest 🌿

In this chapter, we turn to the "D" in ROOTED: Design the Setting—how we shape rooms, tools, and everyday spaces so they help instead of hinder Nature-Informed work.

Most helping spaces have been designed for efficiency, not aliveness. Fluorescent light, stale air, hard surfaces, closed windows, and rooms with no hint of nature all shape what becomes possible inside them. If nature helps regulate, soothe, and reconnect, then the settings we create matter more than we often admit. This chapter explores how rooms, objects, design choices, and even small sensory details can help a space remember the forest:

- how the *illusion of separation* keeps us thinking nature is "out there"

- ways to bring biophilic design into offices, homes, schools, and care facilities

- aquariums, soundscapes, and indoor micro-ecosystems as calming co-therapists

- nature-based metaphors and the Kawa Model as indoor rivers for the psyche

- how technology can become a trailhead toward nature rather than a barrier

- food and cooking as daily, sensory bridges to land and lineage

- how different landscapes act as mirrors for different nervous systems

- a practical exercise to create your own Nature Nook

A Houseplant's Silent Lessons 🌿/💭

Lily never thought of herself as a "plant person." Every plant she brought home eventually withered under her well-meaning but inconsistent care.

Then her grandmother died and left behind a small, drooping pothos.

This one felt different. It wasn't just a plant; it was an heirloom of love and memory. Lily hesitated to take it home, afraid she would fail again. Still, she placed it on her windowsill, unsure but willing.

She watered it.

Adjusted the light.

Checked it each morning.

Slowly, the pothos responded. New tendrils appeared, tender and bright. Leaves unfurled. The vines began to spill over the pot, exploring the air.

One afternoon, misting its leaves, Lily whispered, "You're doing well."
And realized, in the same moment, that she was speaking to herself.

In that small act: caring for a living thing inside her home, Lily discovered an intimate truth: the space we live in can become a quiet ecosystem of mutual care. The room itself can participate in healing.

To design a space well is not simply to arrange furniture or choose pleasing colors. It is to ask what in the room affirms that life is still growing, and how the space meets the body and spirit before language ever begins.

The Illusion of Separation 🌿

For a long time, I believed "real nature" meant being outside: boots on a trail, wind on my face, river spray on my skin.

Then I watched David Attenborough's *Planet Earth*.

I was in my living room, wrapped in a blanket, surrounded by ordinary walls. Yet as snow leopards leapt across cliffs and deep-sea creatures glowed on the screen, I felt my chest ache with awe.

I was not in the jungle or on the reef. But something in me recognized these places as part of my own story. The series did more than inform me; it reawakened belonging.

That night, I understood more clearly that nature is not confined to outdoor adventures; it reaches us in many forms—through images, stories, sounds, objects, food, and memory. The line between what we think of as "inside" and "outside" is far thinner than it appears, and the natural world has countless ways of finding its way into our lives.

The illusion of separation says: *"Nature is out there. This room is separate."*
Nature-Informed design replies: *"The room is part of the watershed. Let's help it remember."*

Biophilic Design – Let the Room Help You 🌿/🐾

When I redesigned my counseling office, I started with a simple question: "How can I make this room feel like it's in conversation with the forest outside?"

I didn't have a skylight or a waterfall. I had four walls and a limited budget—so I began with what I could change.

A large forest photograph filled one wall, creating the sense of stepping into a grove. I added recycled pallet-wood furniture—rough edges, visible grain, a hint of tree still present. A basket of nature objects—stones, pinecones, dried leaves—sat within reach, something to hold and explore. I kept the shades open; natural light, even filtered and imperfect, was invited in. In the waiting room, a small screen looped images and gentle sounds of water.

Chapter 11: D — Design the Setting: Let the Room **Remember** the **Forest**

The space felt different. A person commented, "I feel like I can breathe deeper in here."

This is the heart of biophilic design: reintroducing the elements our bodies evolved with—light that shifts over the day, plants that grow and respond, materials that carry the memory of earth (wood, stone, natural fibers), textures that aren't all plastic and metal.

You don't need a designer's budget. Small shifts can matter: one plant on a desk; a wooden bowl instead of plastic; a nature print where the eyes can rest between clients; a window cleared of clutter so the sky is visible.

Ask of each room you work in:

- "If this space could whisper to the nervous system, what would it say?"
- "How can I help it say: 'You belong to a living world'?"

Aquariums – Water You Can Sit Beside

In one therapy office I visited, the waiting room had a small, unremarkable aquarium. I sat down, distracted by my to-do list, and then found my gaze drawn to the fish.

They drifted, turned, and glided in slow, looping paths. Bubbles rose in a steady stream. Light shifted across gravel and plants. Within minutes, my shoulders softened, my breath slowed, and my thoughts quieted without effort.

Aquariums are worlds in miniature. They reflect ecosystem balance—plants, fish, rocks, snails, bacteria—while offering gentle movement for tired eyes and bringing the element of water into indoor space. Research suggests that even brief time watching fish in an aquarium can lower blood pressure, reduce anxiety, and improve mood. For many people, water is a natural regulator, even when it arrives in a glass box.

174

Practical tips:

- choose an aquarium size you can realistically maintain

- prioritize clear water, simple decoration, and healthy fish over elaborate displays

- place it where people can watch without feeling "on display" themselves

For clients, an aquarium can become a grounding focal point in a waiting room, a metaphor for boundaries ("this is a contained, cared-for world"), or a sensory anchor for guided imagery ("imagine your breath moving like these fish").

Water remembers our origins. Even a small tank can help bodies recall a calmer tide.

Nature Is for Everyone – Nursing Home Landscapes Indoors 🌿/🪴

Many elders cannot stroll in the woods or stand at an ocean's edge. Mobility limitations, health conditions, and institutional routines keep them mostly indoors.

Nature-Informed Care asks: How do we bring the forest to them?

In one nursing-home collaboration, we tried small, intentional interventions. Bird feeders were mounted near windows so residents could watch chickadees, cardinals, and finches up close. An indoor herb station invited touch and smell—basil, mint, rosemary, lavender—awakening memories of gardens and kitchens. Soundscapes of water, wind, and birdsong played softly in common rooms. Screens displayed rotating images of forests, oceans, and meadows. Staff facilitated story circles where residents shared favorite nature memories.

Moments like these reminded us how powerful even indirect experiences of nature can be. One person who watched a simple nature walk YouTube video wrote, "Thank you for this video. I am disabled and unable to leave my home, so these types of videos take me to places I would love to walk in. I went to the Hoh Rain Forest as a child and remember it vividly. This was a relaxing and calming way to revisit it." Even when someone cannot physically step into a forest or along a trail, the sights, sounds, and memories of nature can still carry them there—awakening calm, connection, and a sense of place.

We also experimented with grounding through soil: shallow containers of clean soil were set up with chairs nearby, and residents could rest bare feet or hands in the soil, feeling its cool, grainy texture.

Short-term contact like this can support skin microbiota diversity and may carry immune and mood benefits. But beyond physiology, the expression on residents' faces said enough: "I missed feeling the earth like this."

Even inside a care facility, these small gestures reduced agitation, sparked conversation, and rekindled a sense of aliveness.

Nature is not a luxury for the able-bodied. It is a birthright—and our settings often decide who gets access.

After one of my talks on rewilding the human psyche, a woman came up to share a story that has stayed with me. Her dear friend was dying in hospice. She had been unconscious for some time, and the staff kept saying it would likely be "any moment now," yet the moment did not come. It seemed as though her friend was somehow holding on. One afternoon, while sitting beside her bed and sharing memories of the years they had spent gardening together, a thought came to her. She moved the bed closer to the window, pulled back the curtains, and opened it. A soft draft entered the room along with the sound of birdsong. Within about ten minutes, her friend died. The woman told me it felt as if something changed in the room the moment the barrier between the building and the living world outside was removed as though the breeze itself had carried her friend's spirit into freedom.

Metaphors – How Nature Sneaks into Our Sentences 🌿/🌵

Even when someone has not been in a forest for decades, nature often shows up in their language.

We say:

- "I'm weathering a storm."

- "I'm at a crossroads."

- "We planted the seed, now we wait."

- "I feel like I'm drowning."

- "Something new is blooming in me."

Nature metaphors are not just pretty phrases. They are maps the psyche chooses.

Ecopsychologist Ralph Metzner suggested that what we notice in nature—and the images we instinctively reach for—often mirror our inner world. The river we describe outside frequently reveals the river running through us. One client once described her grief as "a river after heavy rain—rushing, overflowing, muddy and wild."

Months later, after much therapeutic work, she said, "It's still a river. But now it feels more like a steady stream. It still flows, but it's not washing me away."

The landscape never changed; her relationship with it did.

We can invite this kind of metaphor work with simple questions such as, "If your grief were a place in nature, what would it look like?" or "If your resilience were a tree or landform, what might it be?" Through reflections like these, nature begins to function in multiple roles at once serving as a mirror that reflects inner emotional states, a mentor that reveals how living systems adapt and recover, and a kind of medicine that offers images capable of soothing, orienting, and guiding us through difficult moments.

The Kawa Model – Reading the River of a Life 🪴/🌿

The Kawa Model, developed in occupational therapy, invites people to imagine their lives as rivers. The word kawa comes from Japanese, meaning "river," and the name reflects the model's culturally grounded emphasis on relationship, context, and the life flow shaped by both inner and outer conditions.

- The water is their life flow—their energy, purpose, movement.

- Rocks are obstacles—illness, trauma, stressors that impede flow.

- Driftwood represents personal strengths and resources—skills, relationships, faith, creativity.

- The riverbanks are the environment—cultural norms, social context, physical settings.

Often, clients experience relief when they see their struggles as part of a dynamic ecosystem rather than as fixed personal defects. The river teaches that obstruction is not failure; it is part of the terrain. Flow is not about eliminating all rocks, but about widening the space around them.

A river also reminds us that change is constant. Some stretches are narrow and turbulent. Others open into calm expanses. And sometimes, the greatest fear comes not from the rocks—but from the widening.

In his poem Fear, Kahlil Gibran writes:

It is said that before entering the sea

a river trembles with fear.

She looks back at the path she has traveled,

from the peaks of the mountains,

the long winding road crossing forests and villages.

And in front of her,

she sees an ocean so vast,

that to enter

there seems nothing more than to disappear forever.

But there is no other way.

The river cannot go back.

Nobody can go back.

To go back is impossible in existence.

The river needs to take the risk

of entering the ocean

because only then will fear disappear,

because that's where the river will know

it's not about disappearing into the ocean,

but of becoming the ocean.

This poem deepens the metaphor. Sometimes the narrowing is not the problem. Sometimes the fear arises at the threshold of expansion— retirement, an empty nest, recovery, forgiveness, a new identity after loss.

The river does not disappear. It becomes something larger.

The Kawa Model reminds us that we don't need to be outdoors to use nature's patterns as guides. Rivers live inside us—flowing, constricting, widening, trembling, and ultimately moving toward something greater.

Using the Kawa Model indoors is still nature-informed; it uses a river to organize experience.

You might guide someone through:

1. **Draw Your River**

> o Invite them to sketch their life as a river, no artistic talent needed.

2. **Name the Rocks**

> o What blocks flow right now? (grief, caregiving load, debt, illness, burnout)

3. **Find the Driftwood**

> o What supports are present? (friends, humor, therapy, nature time, spirituality, pets)

4. **Study the Riverbanks**

> o How does the environment help or hinder? (neighborhood, culture, work setting, access to green space)

5. **Consider the Flow**

> o What small shifts—moving a rock, adding driftwood, reshaping a bank—could help the water move more freely?

Clients often see their situation differently once it's visualized as a river. Problems become part of a dynamic system rather than fixed, internal defects.

The Kawa Model is a reminder: we don't need to be outdoors to use nature's patterns as guides.

Landscapes as Mirrors

In trainings, I sometimes show a sequence of landscapes—desert, forest, mountain, ocean—and ask participants, "What words or feelings arise for you with each image?"

The answers are wildly diverse:

- Desert: "empty," "lonely," "harsh" for some; "spacious," "pure," "simple" for others.

- Forest: "safe," "mysterious," "cozy" for some; "claustrophobic," "dark," "overwhelming" for others.

- Ocean: "free," "expansive," "healing" for some; "terrifying," "chaotic," "out of control" for others.

The landscapes are the same. The responses are not.

That contrast reveals how personal history, ancestry, and culture shape our felt sense of place. It also explains why one client settles in dense woods while another relaxes only in open fields—and why choosing a setting for outdoor work (or indoor imagery) is never just logistical. It's clinical.

Questions you can explore include "Which landscapes feel like home to you?", "Which feel foreign or unsettling?", and "Is there a landscape you're curious about, even if it scares you a little?" Even urban nature—trees between buildings, a single bird on a windowsill, a strip of sky between rooftops—can act as a landscape mirror. The key is noticing not just what people see, but how it lands in their nervous system.

Befriending Technology – Screens as Trailheads, Not Traps

It's easy to blame technology for our disconnection from nature. And often, that blame is earned.

Yet used wisely, technology can become a bridge.

My relationship with birds changed when I discovered the Merlin Bird ID app. At first, I was only mildly curious about the different songs in the trees. Then I started recording them. The app would name what I was hearing: Northern cardinal. Carolina wren. Song sparrow.

Suddenly, the anonymous chorus became a neighborhood of specific beings. Walks began to feel like conversation: *Oh—you're back this year. I didn't know you lived here, too.*

Chapter 11: D — Design the Setting: Let the Room **Remember** the **Forest**

Now I sometimes introduce Merlin to nature therapy groups. Participants are surprised by how many birds share space with us, even in ordinary places. Something shifts from *alone* to *accompanied*.

Other tools can serve a similar purpose:

- VR nature experiences: brief virtual "forest" or "ocean" immersions have been shown to reduce negative mood and increase nature connectedness, especially for those who can't easily go outside.

- Photo scavenger hunts: using a phone camera to capture "five small, beautiful things" in a parking lot or hospital campus can rewild attention.

- Identification and prompt apps: Tools like Seek (and similar guided nature apps) can offer identification, micro-practices, and simple prompts to notice.

One client who struggled with traditional breath-focused meditation found that a ten-minute video of rustling leaves and flowing water finally allowed their body to soften. Nature, filtered through a screen, reached them in a way bare instructions had not.

The question is not "technology or nature?" but: "How can this tool help someone notice, name, and nurture their relationship with Nature?"

Food & Cooking – Eating as Daily Ecology

Every bite we take is a form of relationship with the Earth.

Soil, rain, sunlight, microbes, pollinators, farmers, transport, markets—so much converges in a single mouthful. And yet, in daily life, food can easily be reduced to calories, macros, or whatever we can reach for between meetings.

When we slow down, food begins to return to its fuller meaning. It becomes a sensory encounter with the living world—color, texture, temperature, aroma, taste. It becomes a story of place, shaped by season and distance, by what is grown nearby and what has traveled far to reach us. It becomes a story of lineage as well, carried through recipes, herbs, and ways of preparing nourishment that have been passed from one generation to the next.

One social worker shared that her grandparents were the ones who first taught her about medicinal herbs. As a child, she learned which plants could calm a racing heart, which teas might soothe a restless mind, and how to gather respectfully and prepare simple remedies with care.

Now, in an urban counseling center, she keeps a small apothecary corner. With clients, she invites them to smell and choose herbs according to what they need in the moment. She teaches simple tea blends for calm, focus, or comfort, and when it feels appropriate, she helps connect these practices back to cultural memory and family tradition. In her work, kettles and mugs have become quiet companions in the healing process.

Food and drink can also become places of reciprocity. We give thanks before eating. We choose, when we are able, to support local growers and ethical producers. We pay attention to waste. We learn where ingredients come from, how they were grown, and what relationships made them possible.

Then a meal becomes more than fuel. It becomes a quiet reminder: *You are made in relationship—with land, with water, and with many human and more-than-human hands.*

Practice: For All Practitioners

Create a Nature Nook

Purpose

To carve out a small indoor sanctuary where nature can meet you daily—through objects, light, and memory—even when you can't go far. You can use this for yourself, with clients, or as a group activity (each person designing their own nook at home or work).

Chapter 11: D — Design the Setting: Let the Room **Remember** the Forest 🌿

Step 1: Choose Your Corner

Look for a spot that can be yours, even if it's tiny:

- a windowsill
- a small shelf
- a bedside table
- a corner of a desk
- a chair by a radiator with a little ledge nearby

The goal is not size; it's intention.

Step 2: Invite Natural Elements

Gather a few items that bring the natural world into that space:

- a houseplant or vase of branches
- stones, shells, pinecones, or driftwood
- a small bowl of soil or sand
- a nature photograph or painting
- a piece of wood, woven basket, or natural fabric

Choose things that feel good to your eyes and hands, not just things that "look Instagram-worthy."

Step 3: Engage the Senses

Now ask yourself: What could this nook smell like? Sound like? Feel like?

In my own office, I keep a small clay bowl of dried lavender and rosemary that I can crush between my fingers when I need a moment of grounding between sessions. A colleague of mine plays a quiet loop of rain sounds from a small speaker tucked behind a plant — her clients often don't consciously notice it, but their shoulders drop within minutes of sitting down. Another practitioner I know keeps a smooth river stone on her desk, cool and heavy, something to hold during hard conversations.

You might consider:

184

- a beeswax or soy candle (unscented or mildly scented)
- a small bowl of herbs (lavender, mint, rosemary) to crush and smell
- a speaker or simple device to play gentle nature sounds
- a smooth stone to hold, a soft blanket, or a textured mat underfoot

The aim is a small multisensory experience: "a pocket of forest" inside your life.

Step 4: Make It Personal

Add 1–3 items that carry meaning:

- a photo of a place you love outdoors
- a journal for nature reflections or sketches
- a gift from someone dear—a feather, pressed leaf, or postcard
- a poem or quote that reminds you of your bond with the Earth

Think of this as your indoor altar to connection, secular or spiritual, simple or elaborate, but *yours*.

Step 5: Start a Collection of Nature Memories

Create a small box, board, or notebook for:

- printed photos of hikes, parks, or gardens that have supported you
- small found objects (a stone from a hard but meaningful day, a leaf from a special walk)
- written memories: three-line stories of "a time nature helped me"

Let this collection grow over months. It becomes a visual map of your relationship with the living world.

Step 6: Use It Daily

Choose a brief, repeatable ritual:

- three slow breaths in your nature nook each morning

Chapter 11: D — Design the Setting: Let the Room **Remember** the Forest 🌿

- one cup of tea sipped there in silence

- five minutes of journaling or sketching once a day

- touching one item and silently saying, "Thank you," before bed

When life becomes overwhelming, this nook is where you can go to remember: *"I belong to a wider, breathing world."*

Chapter Highlights — What to Carry Forward 🌿

- Nature is not limited to outdoor spaces. Once we see through the illusion of separation, we can recognize and invite nature into every room we inhabit.

- Biophilic design—using light, plants, natural materials, and images—helps rooms support regulation and connection. Even small changes can alter how a nervous system feels in a space.

- Micro-ecosystems like aquariums remind us of water's calming presence and can lower stress and anxiety, offering a focal point for rest and contemplation.

- Nature-Informed Care is for everyone, including elders and others with limited mobility. Bird feeders, soil contact, herb stations, soundscapes, and nature imagery can all bring the outdoors in.

- Nature-based metaphors and models like Kawa turn landscapes into mirrors for the inner life, helping people see their challenges and strengths as part of a flowing system rather than fixed traits.

- Different landscapes evoke different emotional responses; attending to these responses helps us choose settings—physical and imaginal—that fit each person's needs and history.

- Technology can be a trailhead rather than a trap: bird ID apps, VR nature, nature videos, and photo scavenger hunts can deepen connection when used intentionally.

- Food and cooking are daily points of contact with earth, season, and ancestry. Herbs, teas, and mindful meals can make Nature-Informed Care tangible even in urban contexts.

- Creating a small Nature Nook is a simple, powerful practice: a personal corner where light, objects, and memory conspire to remind us that we are always living in a larger, living world.

Designing the setting is more than décor. It is about letting every space—office, classroom, bedroom, nursing home hallway—whisper the same truth: *You are not cut off. You are part of a living web, even here, even now.*

Nature Informed Therapy Session Structure

01

Orientation toward land (may include land ack.)

"I'd like to take a moment to tell you a bit about the story of this land we are on ...." History, previous stewards of the land, any trauma of the land.

02

Intro of Co-therapist nature

"I'd also like to introduce my co-therapist for today. Nature has built-in mental health properties. We make a great team. This environment right here has a multitude of therapeutic effects on your physical and mental well-being."

03

Possible intentional nature pause

Inform your client in the beginning of session that you may interject a pause in the middle of session to help recenter and facilitate clarity. "As promised, I am going to pause our conversation for a bit so we can move out of the head and into the body. Nature is a great alley for present moment awareness. Let's start with our sense of touch..."

04

Closing: Gratitude & Reciprocity

"Our session is coming to an end and I want to give us the opportunity to express our gratitude to nature for the healing provided. How would you like to express gratitude today?"

Part III — Application & Leadership

In Part III, we focus on weaving Nature-Informed work into daily life and system-designing settings, tending sustainability, and carrying this practice forward in a way that is both ethical and enduring

Chapter 12: Outdoor Practice, Safety & Groups: Brave Enough, Safe Enough 🌿

In this chapter, we turn to the gritty, practical side of Nature-Informed work: what it actually takes to step outside with people and stay "safe enough" to grow.

We'll look at:

- how to read the land and weather so nature becomes teacher, not ambush

- ways to build psychological safety and trust in outdoor groups

- core outdoor leadership skills every Nature-Informed Practitioner should know

- safety and risk management that honors both courage and limits

- how cultural histories and lived experience shape comfort outdoors

- practical packing and gear guidelines for different types of outings

- working with seasons, weather, and conditions instead of fighting them

- tending your own limits so you don't burn out in the elements

- weaving reciprocity and Leave No Trace into everyday practice

- a planning exercise to prepare yourself and your clients for real-world weather

"I Came Back Different" 🌿/💭

"When the storm finally passed, I realized something: you can't control the weather life throws at you—but you can learn how to stand in it, and sometimes those storms leave you stronger than you were before."

— Participant from a Grief and Backpacking Trip

Storms have a way of rearranging us.

We go out thinking we're in charge: we checked the forecast, packed the snacks, planned the route. Then the sky darkens, the air turns electric, and we remember—very quickly—who has the final say.

Nature is breathtaking. Nature is also indifferent to our schedules.

Outdoor Nature-Informed work lives in this tension. Our task is not to eliminate risk but to befriend it: to prepare, to adapt, to learn when to lean in and when to turn back. The goal is not perfect safety, that doesn't exist, but brave enough, safe enough.

Adapting to the Environment – Reading the Sky, Trusting the Gut 🌿/🐾

I remember hiking with my kids in Wyoming, all of us carrying the same dream: reach the saddle of Grand Teton.

They were strong, eager rock climbers. The trail pulled us higher and higher, past 10,000 feet, into a landscape that felt almost lunar—bare rock, thin air, no shelter.

Then the sky changed.

The light went strange, like a filter over the world. The air thickened. In another life I might have ignored it, but years outdoors had taught me to listen when the wind shifts tone.

Within minutes, the storm arrived.

Lightning cracked around us, each flash too close. The wind roared so fiercely we had to crouch. Rain came sideways. We scrambled into lightning position, bodies hunched low, hearts pounding.

It was not romantic. It was terrifying.

When the storm finally moved past and the world quieted, the tension eased just enough in me. I had renewed respect for how quickly conditions can change. I understood—viscerally—why prevention matters more than bravado. And I felt strangely grateful: for survival, and for the reminder that the mountain sets the terms.

Since then, "adapting to the environment" has meant more than checking a weather app. It means watching the sky and listening for subtle cues (wind direction, cloud build, a sudden temperature drop), choosing routes with bail-out options instead of "no way back but through," and being willing to change the plan mid-session when terrain or weather shifts.

As practitioners, we can treat the environment as a co-facilitator: terrain affects pace, accessibility, and felt safety; weather shapes mood and metaphor—haze, fog, bright sun, mist, snow; and season influences everything from clothing to emotional themes. Good outdoor work begins with curiosity and caution, asking: *What is this place asking of us today?* At the same time, we carry a responsibility for the well-being of those who trust us to guide them. In outdoor settings, that often means choosing safety over convenience. I even had to turn someone away from a hiking day trip when they arrived wearing sneakers instead of appropriate footwear. It is never an easy moment, but part of practicing responsibly in nature is remembering that "better safe than sorry" is not just a saying—it is a commitment to the care of the people in our charge.

Managing Group Dynamics Outside – Circles, Colors, and Choice 🌿/🌱

Groups often begin to move differently under open sky than they do beneath fluorescent lights.

Before a hike, a grief walk, or a day of backpacking, we often begin in the same simple way: standing or sitting together in a circle. No desks. No podium. Just people meeting one another at eye level.

From there, we invite each person to name one outdoor worry. It may be fear of thunderstorms, discomfort with the dark, anxiety about not keeping up, or the unease of sleeping in a tent. We do not rush to fix these fears or talk anyone out of them. We let them be spoken. We let them be heard.

Amy Edmondson describes psychological safety as the belief that one can speak up with questions, concerns, ideas, or mistakes without being punished or humiliated. Outdoors, that kind of safety takes on a very embodied form. It means being able to say, *I'm cold. I'm scared. I'm tired. This feels like too much for me.* It means naming overwhelm without shame. It means being free to choose a smaller challenge without losing dignity in the eyes of the group.

A few tools we use often:

- Nature buddies: Everyone has a partner who checks in throughout the day using a simple "green, yellow, or red?"

- Green / Yellow / Red Zones: A shared language:
    - Green: "I'm okay. This is challenging but manageable."
    - Yellow: "I'm stretched. I could use a pause, encouragement, or adjustment."
    - Red: "I'm overwhelmed. I need to stop, change something, or step back."

- Challenge by choice: We invite, never coerce. Participants can sit out an activity, choose a gentler option, or step away.

Evenings often end with council by the fire: we pass a talking piece; each person can speak or pass; the group listens more than it responds.

Day by day, these rituals normalize fear and vulnerability, build trust—not only with leaders but among participants—and turn the wildness around us into a shared classroom, not an ordeal.

Outdoor Leadership – Skills that Let Nature Do the Therapy ♈/🌿

Outdoor leadership is not about being the toughest person on the trail. It's about being the steady one when everything changes.

Many Nature-Informed practitioners deepen their skills through programs such as NOLS (National Outdoor Leadership School) or Outward Bound. These trainings cover navigation and route planning, risk assessment and decision-making under pressure, wilderness first aid and emergency response, and group dynamics in the backcountry.

Those skills matter because outdoor leadership is not about being the toughest person on the trail. It's about being the steady one when everything changes.

We were caught off guard by an early snowstorm. I was cold, tired, and more than a little afraid. And still, we found our way through it— building proper shelter, layering with care, and eventually making a fire and a hot meal, including what remains one of the more legendary improvised backpacking pizzas.

Those days taught me something I have carried ever since. Cold can be deeply uncomfortable without necessarily being dangerous, if it is met with respect. I learned that I could do hard things when I was prepared and supported. And I learned, too, that joy and laughter do not disappear in the presence of challenge. In other cases they become even more necessary.

For clinicians and facilitators, outdoor leadership asks something similar of us. We are called to anticipate not only environmental weather, but emotional weather as well. We need to communicate calmly when plans change, model self-regulation and flexibility, and discern when it is wise to stretch a little further and when it is time to turn back.

The steadier we are, the more freely we can allow nature to do its own therapeutic work.

Safety & Risk Management – "Safe Enough" in a Living World 🩺/🌿

We often say, "We aim for safe enough."

Not "perfectly safe."

Not "no risk."

Safe enough for growth, learning, and dignity.

Good risk management is less about fear and more about respect. It rests on solid training (Wilderness First Aid - WFA or Wilderness First Responder - WFR for group leaders), clear incident forms and protocols (not because we expect trouble, but because we want to respond well and learn), and the unglamorous basics: insurance, permits, and land-use requirements.

You can think of preparation in five simple layers:

1. Pre-Trip Planning
    - Check detailed weather, daylight hours, and terrain.
    - Identify hazards (exposure, water crossings, wildlife, traffic).
    - Have an alternate route or earlier turnaround point.
2. Team Preparedness
    - Ensure at least one leader has WFA/WFR.
    - Make sure everyone knows emergency procedures and communication plans.
3. Essential Gear
    - First aid kit, navigation tools (map, compass, GPS), communication devices.
    - Extra clothing, food, and water beyond what seems "enough."
4. Safety Briefing

- o Review the green/yellow/red zone framework and buddy system.

- o Clarify roles if something goes wrong: who leads, who calls, who stays.

5. Documentation & Insurance

- o Carry emergency contacts and health information.

- o Confirm permits, land-use requirements, and insurance.

We don't share all the backstage details with clients, but they feel it: the difference between a thrown-together outing and a carefully held one.

Being Prepared – Gear as a Form of Care 

I have a colleague who lives and breathes the Boy Scout motto: *Be prepared.*

He understands something simple and enduring about group life: no matter how many reminder emails are sent, someone will forget something important. So he responds not with irritation, but with quiet foresight. Extra water. Extra rain gear. Extra snacks. His private philosophy seems to be this: *if something is going to be forgotten, let it be something I have already made room for in my pack.*

There is a kind of love in that. Preparation, at its best, is not only logistics. It is a form of care offered in advance.

His packing scales with the outing:

- **50-minute walk-and-talk**
  - o small first aid kit
  - o water (plus a little extra) & snacks
  - o light jacket
  - o phone with emergency contacts and weather app

- **Two-hour outdoor program**
  - o everything above plus extra layers

- o   a simple navigation tool
- **Day hike**
  - o   more complete first aid kit
  - o   multi-tool
  - o   extra water
  - o   detailed map
  - o   more substantial food
- **Multi-day backpacking trip**
  - o   route plan shared with someone who stays home
  - o   shelter and warmth systems
  - o   water purification
  - o   sufficient food plus a small emergency buffer
  - o   redundant safety items (extra lighter, headlamp batteries, etc.)

Essential items for outdoor leaders / Nature-Informed Practitioners:

- First aid kit (appropriate to group size and distance from care)
- Water (more than seems necessary)
- Weather-appropriate clothing (layers, waterproof outer shell; avoid cotton)
- Snacks (trail mix, bars, something salty and something sweet)
- Navigation (map + compass; GPS if appropriate)
- Communication (charged phone, battery pack, sometimes radios)
- Multi-tool or knife
- Emergency shelter (space blanket, bivy, or tarp)
- Light (headlamp + extra batteries)

- Personal meds and basic meds for the group (per local regulations)

- Minimal documentation (emergency contacts, key medical info, incident forms)

Therapy-specific items:

- A weather-resistant journal and pen

- Nature meditation or reflection cards

- A small book or pages of quotes about nature, resilience, grief, growth

Preparedness isn't about paranoia. It's about making room for presence. When leaders know they can handle minor crises, they are freer to notice hawks, shadows, and tears.

Seasons & Weather – Working with What the Sky Offers 🌿/☂

Each season brings its own curriculum. Winter teaches stillness, endurance, grief, and the quiet labor taking place beneath what appears dormant. Spring offers the fragile courage of beginning again. Summer opens toward energy, play, connection, and ripening. Autumn teaches release, harvest, change, and the bittersweetness of endings.

The Center for Nature Informed Therapy came into being just as the pandemic arrived. Almost overnight, indoor sessions became impossible, and we found ourselves carrying the work outside. Clients gathered on a porch wrapped in blankets. I brought tea in a thermos. Heaters and a small fire kept just enough warmth in our hands to stay present with one another.

Some days were brutally cold. Yet people came—often saying, "I'd rather be here than on a screen," or, "This feels more human, somehow." We were returning to an ancient template: people talking about hard things near a fire, under blankets, with the wind reminding us that we are alive.

Weather also offers metaphor: light rain as cleansing, bright sun as energy (and sometimes pressure), fog as "in-between" time, snow as quiet pause—the world under a soft blanket.

One client with pronounced Seasonal Affective Disorder struggled every winter. We shifted some sessions outdoors during daylight hours, explored the Danish idea of hygge (coziness, warm light, ritual comfort), designed a "nature nook" at home, and set small, realistic goals: daily natural light and thirty minutes of movement. SAD didn't vanish, but winter became less of an enemy and more a season to befriend carefully.

Working with weather means planning practically (allergies, heat, cold, ice), letting conditions become part of the therapeutic story—and knowing when to step inside, shorten, or cancel, because no insight is worth frostbite or heatstroke.

Avoiding Burnout When Working in Outdoor Environments

The outdoors is not only physically demanding; it is sensory.

Wind, cold, heat, uneven ground, noise, bugs—all of these land on your nervous system as well as your clients'.

I learned this the hard way on a bitter winter day with five back-to-back 50-minute outdoor sessions. Each client had a manageable dose of cold. I had nearly five hours.

By the end, my body was exhausted. At home, I went straight into a hot bath just to feel my feet again. My muscles ached not just from walking, but from clenching against the cold.

That day taught me I need limits on how many outdoor sessions I do in extreme conditions. It also clarified a simple truth: because storms spike my own fear response, I shouldn't schedule sessions in wooded areas when thunderstorms are forecast. And modeling boundaries— "I'm not able to stay regulated out there in that weather"—is part of ethical practice.

Burnout prevention outdoors looks like building buffers between sessions to warm up, rehydrate, and eat; varying locations and intensities (not every session at maximum exposure); naming your vulnerabilities (dark, storms, steep terrain); using supervision or peer consultation to debrief hard outings; and making sure some time in nature is for you alone—not always for work.

Reflection prompts you might use for yourself:

- When have I felt drained or overextended by outdoor work? What were the early warning signs?

- How do I currently separate "nature for me" from "nature for clients"? Do I need clearer boundaries?

- Where do my own fears or discomforts show up outdoors, and how can I plan around them rather than override them?

You are part of the ecosystem. Your well-being affects the quality of care you offer.

Reciprocity in Action – Leave No Trace, Take 3 for the Sea 🌿

Leave No Trace gives us familiar guidelines:

- plan ahead
- travel and camp on durable surfaces
- dispose of waste properly
- leave what you find
- minimize campfire impacts
- respect wildlife
- be considerate of other visitors

Nature-Informed work adds one more question: *Beyond not harming, how do we give back?*

One small practice I've adopted is inspired by Take 3 for the Sea movement, which encourages people to pick up three pieces of litter whenever they visit the coast to help prevent ocean pollution. I've adapted the idea for forests, parks, and trails. Each time I walk, hike, or meet with a client outdoors, I pick up at least three pieces of litter. I intentionally stick to three, so the practice remains simple and manageable rather than overwhelming. It takes only seconds, but over time it helps keep places cleaner, reminds me that I am a participant rather than a tourist, and gives clients a simple, tangible way to practice reciprocity with the places that support their healing.

You might also volunteer for trail work or habitat restoration, support land trusts or Indigenous-led land-back efforts or build a brief "gratitude to the land" moment into group closings.

Therapy in nature is not only about what Earth can do for us. It's a relationship—and relationships ask for attention, care, and repair.

Practice: For All Practitioners 🌿/☁

Weather Adaptation & Comfort Planning

Purpose

To help you (and, if you choose, your clients) think ahead about weather, comfort, and meaning so that outdoor sessions are less about survival and more about growth.

You can do this as a worksheet for yourself, as part of supervision, or with mentees and trainees.

Step 1: Name Your Common Weather Scenarios

List three weather patterns you encounter most often in your outdoor work, for example:

- Cold + Rainy – damp, bone-chilling, low visibility
- Hot + Humid – heavy air, high sun, risk of dehydration
- Snowy + Windy – low temps, slick surfaces, windchill

You might also add:

- Very windy but dry

- Changeable spring weather (sun, shower, sun again)

Step 2: Build a Gear & Clothing Menu

For each scenario, write what you need to feel reasonably comfortable and safe.

- Cold + Rainy:
    - waterproof jacket and pants
    - insulating mid-layer (fleece, wool)
    - warm hat and gloves
    - wool or synthetic socks (no cotton)
    - waterproof backpack cover or dry bags
    - extra water and snacks

- Hot + Humid:
    - breathable, moisture-wicking clothing
    - wide-brimmed hat, sunglasses
    - plenty of water + electrolytes
    - sunscreen
    - a plan for shade and shorter exposure

- **Snowy + Windy:**
    - insulated jacket, hat, gloves/mittens
    - layered clothing you can adjust
    - hand warmers, thermal blanket
    - extra high-calorie snacks
    - backup heat source for longer outings (if appropriate and safe)

This list becomes your packing template before each season.

As part of this planning, it is also important to consider air quality and pollution, which can affect safety just as much as temperature or precipitation. On days when the Air Quality Index (AQI) is high due to pollution, wildfire smoke, or heavy ozone, outdoor sessions may need to be shortened, moved to shaded or less trafficked areas, or relocated indoors near open windows, plants, or other nature elements. Checking the AQI before heading outside can become part of the same routine as checking the weather.

Step 3: Plan Therapeutic Adjustments & Metaphors

For each weather type, jot down ways you might **lean into** its symbolism:

- Cold + Rainy
    - talk about cleansing, washing away, or the heaviness before renewal
    - explore how people respond to "emotional weather" they can't control
- Hot + Humid
    - connect the intensity of heat with the effort required in inner work
    - notice when everything feels "too much" and explore pacing and shade
- Snowy + Windy
    - reflect on what needs to be protected under the surface
    - consider how people build their own "inner shelter" in hard seasons

These metaphors can be used spontaneously in session or deliberately as prompts.

Step 4: Reflect on Your Own Reactions

Ask yourself how each weather scenario affects your mood and energy, when you feel most resourced outdoors and when you feel frayed or impatient, and what conditions reliably shrink your own window of tolerance. Write down at least one adjustment for each:

- limiting number of sessions in extreme conditions
- choosing different locations (more sheltered, closer to cars, etc.)

- shortening sessions or shifting to hybrid (part outdoors, part indoors/virtual)

Step 5: Model Resilience and Flexibility

Finally, consider:

- How can my preparation itself become a teachable moment for clients?

- What do they learn when I say, "We're going to change the plan because the weather's telling us something important"?

- How can I frame adaptations not as failure, but as wise responsiveness?

Weather becomes not just a background challenge, but a partner in the learning.

Chapter Highlights — What to Carry Forward 🌿

- Outdoor Nature-Informed work lives in a tension between beauty and unpredictability; our goal is not zero risk, but "brave enough, safe enough."

- Adapting to the environment means reading the land and sky, choosing routes with flexibility, and letting nature's cues guide when to proceed and when to change course.

- Group work outside thrives on simple structures: circles, buddy systems, green/yellow/red zones, and evening councils all help create psychological safety amid physical unpredictability.

- Outdoor leadership skills—navigation, first aid, decision-making—allow practitioners to relax enough to notice the therapeutic moments nature offers.

- Safety and risk management are layered practices: training, planning, gear, communication, and clear protocols all work together to support "safe enough" experiences.

Chapter 12: Outdoor Practice, Safety & Groups: Brave Enough, Safe Enough 🌿

- Cultural and historical experiences profoundly shape how people feel in outdoor spaces; adapting locations and plans to honor those realities is essential to equity and trust.

- Thoughtful preparation of gear and supplies is a quiet act of care that frees both leaders and participants to be more present.

- Seasons and weather can be embraced as teachers—each with its own metaphors, physical demands, and invitations to adjust our pace and expectations.

- Practitioners are part of the ecosystem; tending to your own limits and burnout risk is not selfish, it's a prerequisite for sustainable, ethical outdoor work.

- Reciprocity is more than "not harming." Practices like Leave No Trace and simple "take 3" litter rituals embody gratitude and give-back to the places that shelter our work.

When we bring together practical skills, humility, and reciprocity, the outdoors becomes more than a scenic backdrop. It becomes a living partner: sometimes fierce, often generous, always honest, inviting us and our clients into deeper resilience, courage, and connection.

Chapter 13: Designing with Difference in Mind: When Nature Is Not the Same for Everyone 🌿

"You are killing me, fish, the old man thought. But you have a right to. Never have I seen a greater, or more beautiful, or a calmer or more noble thing than you, brother.

Come on and kill me."

— Ernest Hemingway, The Old Man and the Sea

Hemingway's old fisherman knows something we often forget: the forces that feed us can also break us. The ocean sustains him and threatens him in the same breath.

Nature is like that.

It is not universally soothing. It is not experienced the same way by all bodies, all histories, all identities. For some, a forest is sanctuary. For others, it is where danger lives. For some, the sea is medicine. For others, it is the memory of loss.

In this chapter, we'll start by explore how Nature-Informed Therapy (NIT) and Nature-Informed Care (NIC) can honor those differences rather than gloss over them, then move into:

- when nature itself has been the site of trauma or fear
- women's safety and the quiet calculus of risk in outdoor spaces
- tailoring language and framing across cultures and communities
- the "indoor person" identity and what might live beneath it
- climate anxiety and eco-grief—for clients and for us
- nurturing neurodiversity in Nature-Informed work

Hurt by Nature: When the Outdoors Becomes the Wound 🩺

After one of our NIT trainings, a therapist approached me with eyes that contained both excitement and sorrow. She had just completed the program and was ready to bring trees and trails into her clinical work. She could see it clearly: clients walking under green canopies, grief groups gathered at trailheads, nature as co-regulator.

Then a hurricane tore through her community.

The parks closed. Trees lay shattered. Familiar trails became scenes of danger—blocked paths, hanging branches, downed power lines.

"It feels like the land betrayed me," she admitted. "I used to go to the woods to calm down. Now my body tenses just picturing it."

Her story reminded me of Mr. Beaver's line in *The Chronicles of Narnia* about Aslan:

"Safe? Who said anything about safe? 'Course he isn't safe. But he's good."

Nature is often good: beautiful, restorative, generous. But it is not always safe. When storms, wildfires, floods, heat waves, or landslides arrive, the outdoors can become the site of our deepest fear. For some clients, "nature" is not a refuge, it's the source of trauma: tornado sirens and basements, wildfire evacuations and smoke damage, storm surges and neighborhoods gone.

When nature is the source of the wound, we can't offer it lightly as medicine.

With that therapist, our work was not to push her back onto the trail. It was to listen for what she could still trust. The forests were closed, but birds still came to her feeder. The trails no longer felt safe, but morning light still moved across her kitchen floor. Even the sound of rain on the roof—once comforting, now activating—became something we could approach slowly, with care and pacing. Together, we looked for smaller, steadier forms of nature: birdsong, houseplants, a patch of sky through the window, places where the nervous system might begin, little by little, to experiment with safety again.

When someone has been hurt in relationship with nature—through disaster, accident, or a frightening outdoor experience—our role is not to rush repair. It is first to honor the rupture. Of course the body does not trust the woods right now. From there, we move at the pace of trust, beginning with the smallest forms of nature that feel accessible without overwhelming the system. And we make room for both truths at once: that nature can be terrifying, and that it can also be restorative; that it can wound, and that in time, it may also become part of what helps mend.

Being Nature-Informed must include being trauma-informed. That means making room for grief about damaged places, helping people find modest footholds of connection when old ones feel shattered, and knowing where to point people for additional support (for example, disaster behavioral health resources and psychological first aid).

Even after disaster, nature still offers small signals of persistence: a bird returning, a sapling pushing through debris, the first warm day after a long, fearful winter. Those may not erase fear, but they can say: *Something here is still alive with you.*

When a natural disaster ruptures someone's relationship with a place they once loved, healing that rupture may require intentional dialogue— much like the relational repair work used in approaches such as Imago dialogue. "I used to go to the woods to calm down. Now my body tenses just picturing it."

In moments like this, we can have a gentle, relational conversation with the land itself. A practitioner might ask the person to speak first from their own experience: *"Forest, I feel hurt and frightened when I think about what happened. I trusted you to be a place of safety."* Then we invite curiosity: *"If the forest could respond, what might it say?"* Sometimes people imagine the land answering with something like: *"I was wounded too. Storms pass through me just as they pass through you. I am still here, slowly healing."* This kind of imaginative dialogue does not erase the trauma of the event, but it can soften the rupture and reopen the possibility of relationship. In time, people may find their way back to small encounters—a single tree, a patch of grass, a bird on a fence—allowing trust with the natural world to rebuild gradually, one moment of contact at a time.

Women's Safety Outdoors – The Quiet Math of Risk

After a talk on nature and mental health, a woman waited until the crowd thinned and then approached me softly:

"I want to love being outside," she said, "but I never feel safe alone."

She described running in a local park years earlier when a man in a pickup truck began circling, slowing, honking, yelling. He didn't touch her, but he didn't stop. By the time she reached her car, her nervous system had filed an important note:

Out here is not safe for you.

Since then, she rarely goes outdoors without someone else.

Many women I meet carry a similar, invisible load: constant scanning, keys between fingers, phone in hand, route planned—*Can I outrun someone from here?* running quietly in the background.

When I tell people I'm heading off on a solo camping trip, I often hear some version of: "That sounds scary." "I could never do that." "Aren't you afraid?"

Sometimes, yes, I am.

There are nights when a sound splits the dark and adrenaline floods: once a baby fox screamed outside my tent, a sound eerily like a human in distress. My body went rigid. I wanted to pack up and go home. I stayed. When morning came, I felt both humbled and a little stronger.

It is understandable that being alone in nature can stir fear. We are steeped in stories that frame the woods as dangerous, where solitude outdoors is cast as the opening scene of something terrible. And this fear is not simply imagined. For many women, natural spaces can evoke a very real concern about male violence, a concern shaped not only by personal experience, but by culture, media, and broader systems of harm.

At the same time, the larger picture is often more complex than the fear itself suggests. While the perception of danger in wild places may be high, the actual statistical risk of violence there is often lower than people assume, and in many cases lower than in the urban environments through which we move every day. Most women who experience assault are harmed not by a stranger on a trail, but by someone they know.

So we are asked to hold two truths at once. One is the lived reality of fear—shaped by memory, gender, culture, and history. The other is the reality of what is objectively present in the moment. A trauma-informed, nature-informed approach does not use statistics to dismiss fear or persuade someone out of it. It uses them to offer context, while honoring the body's reasons for caution.

As practitioners, we are called first to validate—to recognize that caution often carries its own intelligence. We may then offer context, carefully and only when it is useful, helping fear live alongside a fuller picture of what is actually present. And from there, the work turns toward empowerment: supporting the practices, tools, and choices that help a person feel more able, more resourced, and more fully in command of their own experience.

Practical care can be part of that empowerment. Sometimes it means letting someone know your route and when you expect to return. Sometimes it means choosing busier trails, parks with clear sightlines, or outings that stay close to home. It may involve carrying a charged phone, or in more remote areas, a satellite communicator. For some, walking with a dog or using trekking poles brings an added sense of steadiness. Self-defense training can also help—not as a promise of control, but as a way of inhabiting the body with greater confidence. And for those whose values, circumstances, and local laws support it, carrying legal protective tools may also be part of what helps restore a sense of agency.

We also name that not all bodies feel equally safe outdoors: women, BIPOC folks navigating histories of exclusion and racialized violence, LGBTQ+ people, and people living with disabilities. When we say, "Go out alone and feel free," we have to ask: *For whom is that invitation realistic?* And how can we co-create safety rather than prescribe it?

Emily Dickinson once wrote:

"When much in the Woods as a little Girl, I was told that the Snake would bite me, that I might pick a poisonous flower, or Goblin kidnap me, but I went along and met no one but Angels, who were shyer of me, than I could be of them."

Fear acknowledged can soften. And sometimes, on the other side, there really are angels—or at least foxes and owls, shy deer and rustling leaves—waiting.

Our job is not to push anyone into solitude. Our job is to walk beside them while they decide what kind of relationship with the outdoors feels brave enough and safe enough for their particular story.

Know Your Audience – Tailoring NIT Across Communities ❧/☙

Before a training in Eastern Europe for Ukrainian and Hungarian park rangers, our hosts asked for one change:

"Could you remove the word 'therapy' from your slides?"

They explained that in their region, "therapy" and "psychology" carried heavy stigma. Historically, mental health labels have been used to justify exclusion, institutionalization, or social shame. The word itself closed doors.

So, we adapted.

We shifted from Nature-Informed Therapy to Nature-Informed Care and Well-being. The practice grounding, mindfulness, nature-based regulation, group support—stayed the same. But the language softened, and something in the room relaxed.

It was a vivid reminder that words that feel neutral to us may be loaded elsewhere. "Therapy" might open doors in one setting and close them in another. Cultural humility isn't a paragraph in a training manual—it's a posture.

Before offering any NIT/NIC program, we can ask:

- How is mental health talked about here? With stigma? With openness? Not at all?

- What words feel safer: therapy, counseling, support, well-being, resilience, care, skills?

- How does this community already relate to nature? As sacred? As dangerous? As irrelevant? As work?

- What forms of support already exist, elders, mutual aid, spiritual leaders, community healers, and how can nature complement rather than replace them?

Nature-informed work is not something we "deliver" to a community. It is something we co-create with people who already have their own healing traditions, spiritual understandings, and cultural histories with land.

Often the most nature-informed thing we can do is change a single word on a slide.

"I'm an Indoor Person" – Listening Beneath the Label 🌿/🌵

In consultation with a group of Nature-Informed Therapists, someone asked:

"What do we do when a client says, 'I'm just an indoor person'?"

Chapter 13: Designing with Difference in Mind: When Nature Is Not the **Same** for Everyone

We all laughed knowingly. Then we got quiet.

What does that phrase really mean?

For some people the phrase comes lightly: *I hate bugs. I don't like getting sweaty. I prefer climate control.* But often, when someone says, *I'm an indoor person*, the words are carrying more than preference.

They may be carrying a nervous system shaped by outdoor experiences that felt chaotic, unsafe, or overwhelming. They may be speaking from a body that has never felt fully welcomed in mainstream outdoor culture. Sensory sensitivities may be part of the story too—the heat, the noise, the smells, the textures that become too much too quickly. And for some, the hesitation is rooted in something even more serious: identities that have learned, for good reason, that public outdoor spaces are not always safe.

So before we reach for easy solutions—a plant in the corner, a nature poster on the wall—we pause. We ask what *indoor person* actually means to this particular person. We wonder what has shaped the experience of being outside: memory, culture, health, mobility, identity. We listen for whether there is any part of them that longs for something different, or whether indoors is, in fact, where the deepest sense of ease still lives.

At the same time, it can be helpful to loosen the false divide that so often sits underneath this language. There are not really "nature people" and "non-nature people." There are only humans—each shaped by place, each carrying a relationship with the living world, whether that relationship feels close, distant, conflicted, or forgotten.

One woman who grew up in Beijing told me that silence unsettles her. She sleeps best with the hum of a city, footsteps, voices, horns in the distance. The forest, with its sparse sounds and spaciousness, felt disorienting, not calming. For her, the city was the ecosystem that formed her nervous system.

Another client, a Jamaican man, also described himself as an indoor person. And yet, his whole presence changed when he spoke about cooking. For him, nature lived in the kitchen—in thyme and scallions, in the sound of oil meeting the pan, in recipes handed down as acts of love. His deepest relationship with land was carried through flavor, memory, and shared meals.

So when someone says, *I'm an indoor person,* we listen carefully for what lives beneath the words. Sometimes there is grief there—a quiet longing shaped like *I wish I could enjoy being outside, but…* Sometimes there is relief, a sense that indoors is the one place the body has learned to exhale. And often, both are present at once.

From a nature-informed perspective, connection does not have to begin with wilderness or even with the outdoors in any conventional sense. It may begin on a screened porch, with a potted herb on a windowsill, on a city walk that makes room for street trees and pigeons, or in a busy apartment kitchen filled with steam, spice, and memory.

The invitation is not to become an "outdoorsy person." It is simply this: to notice how nature is already touching your life, and to explore whether there is any gentle way you might want that relationship to deepen.

Climate Anxiety and Eco-Grief – Staying in the Story Without Drowning 🌿/🩺

During a training, a usually quiet participant finally spoke up:

"Does anyone else feel completely overwhelmed by what's happening to the planet?"

The room shifted. Heads nodded. The conversation changed from "how nature heals people" to "how we live while the Earth is hurting."

We, as practitioners, are not outside this story. We read the same headlines: fires, floods, coral bleaching, species loss. We feel the same ache. And we're asked to stay grounded enough to support others.

Joanna Macy, eco-philosopher and Buddhist scholar, writes:

Chapter 13: Designing with Difference in Mind: When Nature Is Not the **Same** for Everyone

"The trouble we are in today is not due to most people being callously indifferent or ignorant, it's because most people fear the pain—so they push grief to the far reaches of the soul."

Eco-anxiety is chronic worry about ecological collapse. It can show up as compulsive news-checking, paralysis in the face of choices ("Does anything I do even matter?"), or dread about having children and imagining the future.

Eco-grief is sorrow in response to actual or anticipated ecological loss: mourning a disappearing coastline, grieving the absence of fireflies that once filled childhood summers, weeping for forests burning in places we've never visited.

These are not signs of pathology. They are signs of *care*.

To be of help, we need to stay within our Window of Tolerance—present, responsive, not numb and not flooded. The Earth doesn't need us cold, and it doesn't need us shattered. It needs us *here*.

One mindset I return to often is:

Allow – Act – Adopt – Attend

- Allow: Let feelings surface: fear, anger, grief, numbness. Give them honest space.

- Act: Take meaningful steps, ideally with others: join a local group, plant trees, support policy efforts, tend one patch of land. Action reduces helplessness.

- Adopt: Choose ways of thinking that hold both urgency and possibility. Avoid both denial and apocalyptic certainty.

- Attend: Care for yourself and others with compassion. Make room for joy, rest, and beauty, not as distractions but as fuel.

Robin Wall Kimmerer reminds us:

"Even a wounded world is feeding us. Even a wounded world holds us, giving us moments of wonder and joy. I choose joy over despair… because joy is what the earth gives me daily and I must return the gift."

A Buddhist monk once told me, when I confessed my grief for the Earth: "Everything is impermanent. Suffering often comes from clinging. Nature changes, and we change with it. Let grief move and keep walking." We are living in a time of profound transition. Letting go is not the same as giving up. Our work is to feel deeply enough to care, and to ground ourselves enough to respond.

Grief counselor Francis Weller reminds us that our sorrow for the Earth also needs expression, inviting communities to create rituals where heartache for the land, the waters, and the losses of our time can be spoken, witnessed, and held together. When grief is given a place to be shared in this way, it no longer isolates us; it becomes part of the collective work of loving and tending the world.

Nurturing Neurodiversity – Nature as a Wider Welcome 🌿/🌵

Nature won't meet every nervous system the same way, and that isn't a problem to fix. It's a richness to honor. At its best, nature is spacious enough to hold many ways of being: the kid who needs to run in circles, the adult who prefers to sit and watch ants, the person who only feels safe on smooth, predictable paths, the one who thrives on climbing rocks and getting muddy.

For neurodivergent individuals (autism, ADHD, dyslexia, sensory processing differences, and more), NIT/NIC can offer sanctuary, but only if we shape it with care.

Here are a few possibilities:

1. **Sensory Gardens & Small Rituals.**

   Planters with soft lamb's ear, fragrant mint, or calming lavender invite touch and smell. Predictable routines help: "First we smell, then we choose, then we brew." One client loved plucking mint leaves for tea each session; the sequence: walk, pick, smell, brew, became a reliable regulating ritual.

2. **Structure & Choice.**

   Open-ended wandering can overwhelm. Many people benefit from a clear frame, such as "Let's find three different textures on this walk," or "Let's search for five shades of green today." Simple, game-like tasks can reduce anxiety and make engagement feel like play rather than performance.

3. **Animals as Gentle Bridges.**

   Nonverbal, nonjudging beings can be easier company than people, watching squirrels, ducks, or birds; visiting a butterfly garden; or having brief, structured contact with therapy animals when appropriate.

4. **Music & Movement.**

   Rhythm regulates. Keep it simple: one activity at a time, with a clear beginning, middle, and end.

5. **Technology as a Friendly Tool.**

   For some clients, unpredictability is the enemy. Identification apps can make nature feel more "knowable," and virtual nature can offer access for those who can't, or don't want to, be outside for long.

Advocate Julie Ayers offers a crucial reminder:

"Stop fighting truths. Some people hate being outside because of bugs. We need to stop trying to get them outside and instead figure out ways to help them connect with nature."

Practice: For All Practitioners

Designing with Difference in Mind

Purpose

This exercise is meant to deepen cultural humility and practice tailoring NIT/NIC for real people with real histories, not idealized "outdoor clients."

Step 1: Choose a Client Profile

Pick one of these—or invent your own based on people you serve:

- a woman recovering from sexual trauma who feels anxious about being alone outdoors

- a neurodivergent teen who finds transitions overwhelming and is easily overloaded

- a refugee client, newly resettled, with no experience of forests or hiking

- a climate activist struggling with despair, burnout, and disillusionment

- an older adult who identifies as an "indoor person" but is curious about gardening

Step 2: Reflect on Context

Journal briefly:

- How might this person's cultural background, neurotype, gender, and lived experience shape their relationship with nature?

- What narratives about outdoors—safety, kinship, danger, class, race—might they have absorbed?

- What parts of "standard" nature programming might feel inaccessible, frightening, or irrelevant to them?

Step 3: Choose a Setting

Imagine a place that could feel safe enough and welcoming for this person:

- a busy urban park bench instead of a secluded trail

- a community garden instead of a wilderness overlook

- a screened porch, balcony, or bright indoor room with plants

- a short, flat path near a parking lot

- a quiet courtyard with a single tree

Name specific qualities:

- Is it open or enclosed?

- Are there other people around or not?

- How easy is it to leave if needed?

Step 4: Design a 20–30 Minute Experience

Create a brief NIT/NIC plan tailored to this person.

Include:

- Setting – where you'll be (indoors/outdoors/hybrid).

- Intention – one simple purpose ("meet this garden," "practice one small calming skill outside," "share climate feelings under this tree").

- One main activity or ritual –

  - "touch and smell three plants,"

  - "walk while naming colors,"

  - "sit and listen for five sounds,"

  - "write a letter to a future Earth,"

  - "water one plant and say thank you."

- Sensory + accessibility considerations –

  - seating, shade, temperature, noise, pathways, bathrooms.

- Closing – a short reflection prompt or grounding:

  - "What, if anything, felt okay out here today?"

  - "Is there one image or moment you'd like to remember?"

Step 5: Optional – Try It Through Their Eyes

If appropriate, go try your designed activity, imagining you are that client:

- What felt good? Awkward? Overwhelming?

- Did the setting feel safe enough?

- What would you change to make it more accessible, dignified, or resonant?

When we design with differences in mind, nature stops being a single doorway everyone is expected to walk through in the same way. It becomes a house with many doors, many thresholds, many rooms, and a potential meeting place between a human nervous system and a living, varied world.

Chapter Highlights — What to Carry Forward 🌿

- **Hurt by Nature** – For some, nature is not calming but terrifying due to disasters or traumatic experiences. NIT/NIC must be trauma-informed, honoring rupture and helping people find small, trustworthy footholds of connection again.

- **Women's Safety Outdoors** – Many women and marginalized folks carry an ongoing risk calculation in outdoor spaces. Their fear is shaped by lived experience and culture, not just statistics. Our work is to validate, contextualize, and co-create realistic safety and agency.

- **Tailoring Across Communities** – Language matters. Shifting from "therapy" to "care" or "well-being" can make Nature-Informed work accessible in settings where mental health terms are stigmatized. Cultural humility means listening first and adapting frameworks, not imposing them.

- **"Indoor Person" Identity** – When clients call themselves "indoor people," it often signals deeper stories—trauma, cultural disconnection, sensory preferences, or identity. Rather than "fixing" them, we get curious about where nature already lives in their lives.

- **Climate Anxiety & Eco-Grief** – Eco-anxiety and eco-grief are not pathologies but signs of care. To be of use, we must stay within our Window of Tolerance, practicing models like Allow–Act–Adopt–Attend and making room for joy as a necessary companion to grief.

- **Neurodiversity & Nature** – NIT/NIC can be profoundly supportive for neurodivergent individuals when experiences are structured, sensory-aware, and choice-full. Sensory gardens, rituals, animals, rhythm, and technology can all serve as gentle bridges.

- **Animals as Teachers** – Animals have long been archetypal guides. In NIT, we approach them with cultural humility, focusing on personal associations and experiences, how it feels, and what it might mirror—rather than appropriating symbolic systems not our own.

Chapter 14: A 30-Day ROOTED™ Practice: From Inspiration to Habit 🌿

On the last afternoon of a 3-day nature therapy training, the energy in the room is usually bright and tender. People are buzzing with ideas, notes are overflowing, and someone will eventually say what many are thinking:

"I'm so inspired. But on Monday the inbox will be full, the schedule packed, and I'm worried I'll fall back into old routines."

This chapter is written for that Monday.

Not to demand a complete overhaul of your work, but to offer a small, steady path—thirty days of simple steps that help ROOTED™ move from the page into your body, your sessions, and your spaces.

You don't have to do it perfectly. You don't have to go outside every day. You are simply invited to take one concrete step at a time toward letting nature become a trusted partner in your care for others.

The 30-day arc roughly follows the book:

- **Week 1 – Remembering:** Your own eco-identity, ethics, access, and safety.

- **Week 2 – R & O:** Regulate Attention and Open the Senses with micro-practices.

- **Week 3 – O & T:** Optimize Dosage and Tend Relationships—with place and the more-than-human world.

- **Week 4 – E & D:** Enact Reciprocity & Meaning and Design the Setting, with an eye toward sustainability.

You can adapt the timing, repeat weeks, or stretch it into two or three months. The structure is a gentle guide, not a rigid prescription.

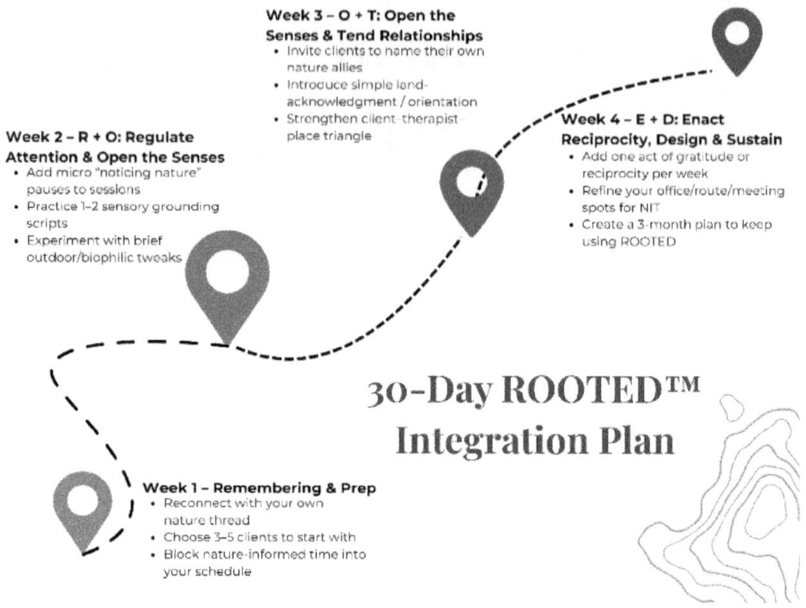

How to Use This 30-Day Plan 🌿/⚕

Choose a small starting circle.

Begin with 3–5 clients who seem curious about nature or already mention it—pets, gardens, walks, longing for the beach, "I miss the park," even "I'm an indoor person but I love my houseplants." You can expand later.

Stay within your scope.

Revisit your setting by considering agency policies, liability and supervision requirements, and access to outdoor or semi-outdoor spaces.

Every ROOTED mechanism can be practiced indoors, in telehealth, or in urban spaces. Outdoor sessions are an option, not a requirement.

Expect wobble.

Some days you'll skip a step. Some weeks you'll feel behind. That's okay. The goal is to lean in, not to "do it right."

Use simple tools.

At the end of this chapter you'll find templates you can adapt:

- a 30-Day Calendar Tracker

- a ROOTED Session Checklist

- a Client Implementation Log

- a Therapist Self-Reflection Sheet

Print them, recreate them in your EHR, or sketch your own versions in a notebook.

Keep an eye on your own nervous system.

As you invite nature into your work, notice what shifts in you: perhaps more breath, more grief, or more spaciousness. Your eco-identity and eco-grief are part of this process. They are not distractions; they are data.

Week 1 (Days 1–7): Remembering & Preparing the Ground

Focus: Part I of the book—Remembering, eco-identity, time, and difference.

Week 1 goals:

- Rekindle *your* relationship with nature as co-therapist.

- Clarify ethics, scope, and safety in your context.

- Identify 3–5 clients to begin with.

- Start simple, honest documentation.

Days 1–2 – Your Eco-Identity & Intention 💬

Revisit Chapters 1–3:

- "Remembering: Nature as Co-Therapist"

- "A Necessary Revolution"

- "Eco Identity"

Chapter 14: A 30-Day ROOTED™ Practice: From Inspiration to Habit 🌱

Complete or refresh the Eco-Autobiography exercise from Chapter 3:

- the landscapes that shaped you

- the places that supported you in hard seasons

- the ruptures and the returns

Then write a short 30-day intention, such as:

"Over the next 30 days, I will invite nature into my work with four clients in ways that are ethical, culturally humble, and grounded in consent. My guiding thread will be: *trees* (or water, sky, gardens…)."

Add this intention to your 30-Day Calendar Tracker.

Days 3–4 – Ethics, Scope & Safety Check 🩺

Return to the ethics and outdoor practice sections of the book.

Make a brief, realistic inventory:

- Places:
  - Which nearby parks, courtyards, or green edges are feasible?
  - What indoor/urban nature options exist—windows, roof decks, atriums, indoor plants, views of sky?

- Policies & coverage:
  - What does your employer or practice policy say about off-site work?
  - What is your liability situation around outdoor sessions or walks?

Draft or refine a simple informed consent paragraph for Nature-Informed work. For example (customize to your setting):

"At times I may invite you to notice or engage with natural elements—such as plants, weather, sounds, or outdoor settings—as part of our work together. You always have the right to decline or modify these invitations. We will choose locations and activities that feel as safe and accessible as possible, and we will regularly check in about your comfort level, boundaries, and identities in relation to these spaces."

Keep this language handy—for your intake paperwork, verbal explanation, or both.

Day 5–6: Choosing Clients & Mapping Threads 🌿

Select 3–5 clients who may benefit from NIT/NIC. For each, jot down:

- presenting concerns
- nature threads already present (dog walking, plants, long drives, love of storms, city park, kitchen herbs, etc.)
- possible barriers (mobility, medical needs, safety concerns, cultural history, eco-anxiety, "hurt by nature," "I'm an indoor person")

Create one row per client in your Client Implementation Log with columns like:

- client initials
- nature thread(s)
- ROOTED focus for the month (e.g., R & O, or O & T)
- safety/consent notes
- cultural / identity considerations

This isn't a research tool. It's a way to see, on paper, that you are not trying to "do nature" with everyone all at once.

Day 7 — Naming the Intention with a Client 🩺

With at least one client this week, speak your intention out loud.

Sample script:

"You've shared how much [the river / your balcony plants / walking your dog] matters to you. Because I practice Nature-Informed Therapy/Care, I'd like to bring a bit more of that connection into our work—only if it feels right for you. It might be as simple as noticing what you see through the window or doing a brief grounding exercise outdoors near the building. How does that idea land with you?"

Note their response in your log, whether it is enthusiastic, cautious, curious, or simply "not now." All of these are useful information. Consent is an ongoing relationship, not a one-time checkbox.

Week 2 (Days 8–14): Regulate Attention & Open the Senses

Focus mechanisms:

- R — Regulate Attention
- O — Open the Senses

These are gentle entry points. They can happen indoors, online, or in the briefest contact with nature.

Week 2 goals:

- Use at least one *R* practice with your chosen clients.
- Use at least one *O (senses)* practice.
- Start noting how their bodies and stories respond.

Days 8–10 – Regulate Attention (R) 🌿

Choose one or two Attention Restoration Theory (ART)-informed practices from Chapter 6 and try them with at least three clients.

Examples:

- Noticing Nature Micro-Reset:
  - even through a window or a photo
  - "Let's spend one minute just letting your eyes rest on something alive or natural."
- Soft Fascination Pause:

o   "Let your gaze land on something in or outside the room that feels gentle to look at. No need to analyze it. Just notice."

In-session script:

"Before we go further into this hard topic, I'd like to give your nervous system a tiny break. Would you be willing to spend sixty seconds just looking out the window, noticing one thing in nature your eyes are drawn to? I'll do it with you. We're not trying to make anything happen—just letting attention rest."

Afterwards, ask, "What do you notice in your body now?" and "Did anything shift in your thoughts or emotions?" Capture a few words in your Client Implementation Log.

Days 11–13 – Open the Senses (O #1) 🌿

From Chapter 7, select *one sensory doorway per client*:

- Smell: a sprig of rosemary, pine, or mint; a drop of essential oil on a cotton ball.

- Sound: brief nature sounds (water, birds, wind) or simply opening a window.

- Touch: a smooth stone, a pinecone, a piece of wood, bare feet on safe ground.

- Sight: plants, a simple nature print, the view outside, an aquarium.

Gentle script:

"Many people find that engaging the senses with something from nature can help them feel present and grounded. Would you be open to trying a very brief sensory element—like a natural scent or a short sound of water—just to see how your body responds?"

Invite description by asking, "What do you notice about this smell, sound, or texture?" and "Do any memories or images come up?" You are supporting both regulation and meaning-making.

Day 14 – Therapist Reflection 💬

Set aside 15–20 minutes for yourself.

On your Therapist Self-Reflection Sheet, consider:

- Which R and O practices felt most natural or alive for me?
- Which felt awkward, forced, or unclear?
- With which client did I see even a small shift in breath, tone, or posture?
- Did any ethical or cultural questions surface?

Choose one small adjustment for the coming week (for example: "Shorten the micro-resets," or "Ask more consent questions before introducing scent").

Week 3 (Days 15–21): Optimize Dosage & Tend Relationships

Focus mechanisms:

- O — Optimize Dosage
- T — Tend Relationships

Now you're looking at how much nature contact makes sense and what kind of relationship clients are building with place and the more-than-human world.

Week 3 goals:

- Create a simple "nature dosage plan" for yourself and 2–3 clients.
- Facilitate at least one intentional "introduction" to a place or being.
- Notice shifts in trust, narrative, and sense of belonging.

Days 15–17 – Optimize Dosage (O #2) 🌿

Return to the Nature Pyramid and self-care plans from Chapter 8.

For yourself:

Look back over the last week:

- daily micro-doses (window views, a plant you water, a five-minute walk)

- weekly medium doses (park time, a longer walk, balcony tea)

- monthly deep doses (half-day hikes, retreats, time by water)

- annual immersions (multi-day trips, pilgrimages, backpacking, or simply repeated visits to a beloved place)

Sketch your next 30 days as a Nature Pyramid and note it on your calendar.

For 2–3 clients:

Ask:

- "On a typical day, how much time do you spend noticing or being in contact with nature—if at all?"

- "Is there a small and realistic way to increase that by even five minutes this week? It could be as simple as sitting by a window or noticing the sky as you walk to your car."

Co-create a tiny, precise step and write it into their plan:

"Between now and next session, would you be willing to spend five minutes, three times this week, sitting by your window and noticing what changes in the light or weather? We'll check in next time about what that was like."

Days 18–20 – Tend Relationships (T) 🌿

Choose one client and one place or being:

- a tree outside your building

- a pond, river, or community garden

- a houseplant in your office

- a patch of sky framed by a window

If you're outside together:

Chapter 14: A 30-Day ROOTED™ Practice: From Inspiration to Habit 🌱

"Before we begin, I'd like to acknowledge this place and thank it for holding us today. You're welcome to do that in your own way. Let's also remember that this land has a long history—of ecosystems and peoples—long before us. We're visitors here."

From there, we might invite participants to notice what they feel drawn toward, and whether the place evokes a sense of welcome, unease, curiosity, or something harder to name. These early impressions often tell us something important about how relationship is beginning.

When we are indoors or online, this kind of relational work can still unfold. A plant, a stone, or even an image of a river or tree can become a point of connection. We might wonder together: *If this tree, this stone, this river could speak, what might it notice about how you show up in its presence? Or, If you were to build a relationship with this place or element over time, what might you hope to receive from it—and what might you hope to offer in return?*

As these conversations deepen, relational metaphors often begin to emerge on their own. A river may become a reminder that movement continues even around obstacles. A thirsty plant may mirror the cost of neglect, or the quiet depletion that comes from giving too much for too long. When these images arise, it can be helpful to capture them. They often become part of the language a person returns to when insight needs to be remembered in a more lived and embodied way.

Day 21 – Midpoint Check & Gentle Evaluation 🩺

Look over your Client Implementation Log:

- Which ROOTED pieces has each person experienced?
- Where have you seen even small shifts in affect, presence, or story?

For each of your 3–5 clients, jot a quick 1–5 rating:

- Presence / regulation
- Sense of connection (to self, others, or place)
- Hope / meaning

This is not data for publication. It's a snapshot to help your intuition and memory.

Week 4 (Days 22–30): Enact Reciprocity, Design the Setting & Sustain

Focus mechanisms:

- E — Enact Reciprocity & Meaning
- D — Design the Setting

Now we braid things together: small acts of giving back, thoughtful spaces, and a realistic plan for life after 30 days.

Week 4 goals:

- Introduce at least one reciprocity or meaning-making practice.
- Make one tangible change to your physical or virtual setting.
- Name a simple, sustainable next phase.

Days 22–24 – Enact Reciprocity & Meaning (E) 🌿

Revisit Chapter 10's reflections on reciprocity, the Honorable Harvest, and awe.

With at least one client, explore an act of giving back that fits their life:

- picking up three pieces of trash at the park after a walk
- watering a shared plant in your office with a brief thank-you
- writing a short note or drawing for a place that's been meaningful ("Dear river / tree / balcony, thank you for…")
- planting herbs or flowers at home or in a community space

Script idea:

"We've been spending time with this place/these plants as they support your healing. Sometimes it can feel grounding to offer something back, even something very small. Is there a way you'd like to thank or care for the natural spaces that have been helping you?"

Emphasize that reciprocity is about relationship, not about "fixing the planet" alone.

Days 25–27 – Design the Setting (D) 🌿

Look around your main working environment—office, telehealth backdrop, classroom, or group room—through the lens of biophilic design from Chapter 11.

Ask:

- "If this room could remind people they belong to a living world, what would need to change?"

- "What is possible within my budget and constraints?"

Choose one concrete change you can make this week:

- adding a plant, branch, or small water feature

- rotating a chair so at least one seat has a view of sky or trees

- simplifying your telehealth backdrop and adding a gentle nature image

- creating a small arrival space with a plant, bowl of stones, or a sensory object from outdoors

Name it with clients:

"You may notice I've brought in some elements from outside. I'm experimenting with how our space can better support nervous system regulation and a sense of connection. I'm curious what, if anything, you notice."

Their responses are part of your learning.

Days 28–29 – Integration Conversations 🩺

With each of your initial 3–5 clients, invite reflection:

- "What has it been like to bring more of nature into our work recently?"

- "Has any part of this surprised or annoyed you?"

- "Are there pieces you'd like to continue, change, or let go of?"

- "Does anything in this approach feel mismatched with your culture, identity, or preferences?"

Listen without defensiveness:

- adjust pace

- adjust language (Nature-Informed *Care* vs *Therapy*)

- adjust setting or sensory load

The client is not just the recipient of NIT/NIC; they are your co-designer.

Day 30 – Your Debrief & the Next Season 💬

Set aside 30–45 minutes, ideally somewhere you can see or feel some bit of nature—a window, a plant, a patch of sky.

On your Therapist Self-Reflection Sheet, explore:

Remembering

- How has my sense of the Earth as co-therapist shifted in these 30 days?

- Do I feel more accompanied in my work?

ROOTED

- Which 1–2 mechanisms came most naturally to integrate?

- Which felt clumsy, intimidating, or still theoretical?

Clients

- Think of one client whose relationship with nature or with themselves changed even a little. What did you notice?

Barriers

- What structural or systemic obstacles did I encounter (time, policies, weather, access, racism, ableism)?

- Where might I need supervision, advocacy, or collaboration?

Next 90 days

Then write:

- One clear commitment for the next three months
    - e.g., "Invite two new clients per month into Nature-Informed practices."
    - "Maintain a weekly 15-minute sit spot for myself."
    - "Talk with my team about liability and access for outdoor sessions."
- One gentle permission
    - "I have permission to go slowly."
    - "I have permission to make mistakes and keep learning."

You are not finishing a program. You are stepping into a **longer conversation** with your clients and the land.

Printable Tools & Trackers (Templates)

You can create these in your own style; here are suggested structures.

1. 30-Day Calendar Tracker

A one-page grid (Days 1–30) with columns such as:

- Date

- ROOTED focus (R, O, O, T, E, D)

- Client initials (or personal practice only)

- Brief note (e.g., "Window noticing, 2 min," "Scent & breath," "Park walk + reciprocity")

2. ROOTED Session Checklist

A simple checklist you can keep on a clipboard or in your EHR:

✓ ☐ Regulate Attention (R) offered?

✓ ☐ Open the Senses (O1) invited?

✓ ☐ Optimize Dosage (O2) discussed or planned?

✓ ☐ Tend Relationships (T) named (to place/more-than-human)?

✓ ☐ Enact Reciprocity/Meaning (E) explored?

✓ ☐ Design the Setting (D) adjusted or noted?

You won't check every box every session; the checklist simply keeps ROOTED visible.

3. Client Implementation Log

Columns might include:

- Client initials
- Date
- ROOTED focus
- Intervention used (1–2 words)
- Client response (felt sense, emotion, quote)
- Follow-up or adjustment needed

4. Therapist Self-Reflection Sheet

A recurring page you can use weekly or monthly:

- Today's date
- How connected do I feel to nature as a co-therapist (1–5)?
- What nature threads showed up in my sessions this week?
- Where did I feel most alive or aligned in my work?
- Where did I feel flat, overextended, or disconnected?
- One small nature-informed step I will take for myself this week.

Chapter Recap 🌿

This 30-day practice is not a test you pass or fail. It is a way to walk ROOTED into your real life:

- You began by remembering your own story with nature and clarifying ethics and safety.

- You experimented with small R and O practices—attention, senses, micro-doses of contact.

- You played with dosage and relationship, letting places and beings become part of the therapeutic field.

- You tried reciprocity and adjusted your settings so they whisper: *You belong to a living world.*

From here, ROOTED is less a checklist and more a way of seeing:

- your office as part of a watershed

- each client as a creature of place and time

- each session as a meeting not just between two humans, but between many forms of life and history

The 30 days are not the end. They are a beginning that you can repeat, stretch, evolve, and share.

Reproducible Worksheets & Session Tools 🌿

30-Day Calendar Tracker

Use this tracker to note small, daily steps you take to integrate ROOTED into your work. You can track your own practices, your client sessions, or both.

Day	Date	ROOTED Focus (R/O/O/T/E/D)	Client(s) / Personal Practice	Notes
1				

2				
3				
4				
5				
6				
7				
8				
9				
10				
11				
12				
13				
14				
15				
16				
17				
18				
19				
20				
21				
22				
23				
24				
25				
26				
27				
28				

29				
30				

ROOTED Session Checklist

Use this checklist as a light-touch reminder of the ROOTED mechanisms during or after sessions. You do not need to check every box every time; it is simply a way to keep the framework visible.

Date / Session ID	Mechanism	Notes (How it showed up in this session)
	R: Regulate Attention	
	O: Open the Senses	
	O: Optimize Dosage	
	T: Tend Relationships	
	E: Enact Reciprocity & Meaning	
	D: Design the Setting	

Client Implementation Log

Use this log to track how you are introducing ROOTED practices with individual clients over time.

Client Initials	Date	ROOTED Focus	Intervention Used	Client Response	Planned Follow-Up

Therapist Self-Reflection Sheet

You can use this reflection sheet weekly or monthly to track how Nature-Informed work is affecting you and your practice.

Date:
How connected do I feel to nature as a co-therapist today (1–5)?
What nature threads showed up in my sessions this week?
Where did I feel most alive or aligned in my work?
Where did I feel flat, overextended, or disconnected?
One small nature-informed step I will take for myself this week:

Chapter 14: A 30-Day ROOTED™ Practice: From Inspiration to Habit

Final Word: A Return, Not a Departure

If you like, close this book for a moment and rest your hand on something natural: the bark of a tree outside, the leaf of a houseplant, a stone on your desk, or even your own chest, rising and falling like a small tide. Take one slow breath.

This work—Nature-Informed Therapy and Nature-Informed Care—is not about adding one more technique to an overcrowded toolbox. It is about remembering something very old:

You have never not been in relationship with nature.

You are made of water, minerals, air. The same elements that move through rivers and soil move through your blood and bones. When you walk into a forest or stand in a parking lot under a winter sky, you are not entering an "environment." You are meeting extended parts of yourself.

Imagine, for a moment, that this remembering spread beyond therapy rooms. Doctors step outside for five minutes with patients between lab results. Teachers begin class with a brief "notice the light" moment at the window. Architects design buildings that let sky and trees into people's daily lives. Custodians, line cooks, parents, chaplains, bus drivers, each finds small, steady ways to weave nature's rhythms into how they care for others.

Imagine a world in which hospitals have gardens that are not decorative but essential; staff meetings happen under trees whenever possible; city planners treat green space as medicine, not luxury; school counselors walk with students under open sky instead of only across metal desks; and it is ordinary, not strange, to cry beneath a tree or pause for one breath with the land before beginning a shift.

Final Word: A Return, Not a Departure

In a time of rising distress and disconnection, we often look for new protocols, new diagnostic labels, new medications. All of these can help. But perhaps part of what we need is ancient: different places, different rhythms, and different relationships—with the Earth and with each other. Not instead of therapy, but alongside it.

Nature-Informed work does not ask you to abandon clinical wisdom. It asks you to expand it: to let wind, water, soil, and birdsong join your team; to consider the nervous system in the context of weather, season, and place; to see each client not as an isolated self, but as a being threaded into a vast, living web.

I'll admit something here: I have always struggled with the phrase *self-care*. Too often it echoes a quiet cultural message that when we are hurting, we should simply fix it ourselves—take another bath, sign up for another yoga class, try harder to manage on our own. But when we are truly struggling, what most of us need is not more isolation disguised as self-care. We need connection. Connection to other people, connection to the natural world, and connection to parts of ourselves that come alive in relationship. Healing rarely happens in a vacuum. It happens in the presence of someone who meets us. The next time you hear the phrase *self-care*, you might pause and think instead about *nature connection*: stepping outside, feeling the air, noticing a tree, or listening to birds. These simple moments of contact remind us that care is not only something we give ourselves, it is something we experience through relationship with the living world around us.

You do not have to start a program, write a grant, or lead backpacking trips to participate in this shift. You can offer a client one mindful look at the sky, place a plant between two chairs, drink tea made from leaves that grew in soil you can picture, pick up a piece of trash on the way back to your car, or sit for five minutes a week in one spot outdoors, letting that place come to know you.

These are small acts. But they accumulate. They change rooms. They change bodies. They change stories.

Nature is already reaching for you, and for the people you serve—in the way light moves across your office wall, in the birds outside the hospital window, in the weeds pushing through sidewalk cracks. Your task is not to manufacture connection, but to notice it, name it, and nurture it.

Thank you for walking this path with me, for your willingness to experiment, to listen to land and body, and to let therapy stretch beyond four walls. May the next time you step outside—onto a trail, onto a balcony, into a parking lot—you feel, even for one breath, that you are being greeted. And may you carry that greeting into your work, your community, and your own life, one rooted step at a time.

References

Barreto, M., & Ellemers, N. (2003). The effects of being categorized: The interplay between internal and external social identities. *European Review of Social Psychology, 14*(1), 139–170. https://doi.org/10.1080/10463280340000045

Bartlett, C., Marshall, M., & Marshall, A. (2012). Two-Eyed Seeing and other lessons learned within a co-learning journey of bringing together Indigenous and mainstream knowledges and ways of knowing. *Journal of Environmental Studies and Sciences, 2*(4), 331–340.

Beatley, T. (2012, August 7). *Exploring the nature pyramid.* The Nature of Cities. https://www.thenatureofcities.com/2012/08/07/exploring-the-nature-pyramid/

Berman, M. (2025). *Nature and the Mind: The Science of How Nature Improves Cognitive, Physical, and Social Well-Being.* S&S/Simon Element.

Berto, R. (2005). Exposure to restorative environments helps restore attentional capacity. *Journal of Environmental Psychology, 25*(3), 249–259. https://doi.org/10.1016/j.jenvp.2005.07.001

Bird, W. (2007). *Natural thinking: Investigating the links between the natural environment, biodiversity and mental health.* Royal Society for the Protection of Birds.

References

Bowlby, J. (1988). *A secure base: Parent-child attachment and healthy human development.* Basic Books.

Bratman, G. N., Daily, G. C., Levy, B. J., & Gross, J. J. (2015). The benefits of nature experience: Improved affect and cognition. *Landscape and Urban Planning, 138,* 41–50. https://doi.org/10.1016/j.landurbplan.2015.02.005

Caron, C. (2024, February 5). Therapists trade the couch for the great outdoors. *The New York Times.* https://www.nytimes.com/2024/02/05/well/mind/outdoor-therapy-depression-anxiety.html

Edmondson, A. (1999). Psychological safety and learning behavior in work teams. *Administrative Science Quarterly, 44*(2), 350–383. https://doi.org/10.2307/2666999

Faber Taylor, A., & Kuo, F. E. (2006). Is contact with nature important for healthy child development? State of the evidence. In C. Spencer & M. Blades (Eds.), *Children and their environments: Learning, using and designing spaces* (pp. 124–140). Cambridge University Press.

Four Arrows (Wahinkpe Topa), & Narvaez, D. (2022). *Restoring the kinship worldview: Indigenous voices introduce 28 precepts for rebalancing life on planet Earth.* North Atlantic Books.

Frantz, C., & Mayer, F. S. (2014). The importance of connection to nature in assessing environmental education programs. *Studies in Educational Evaluation, 41,* 85–89. https://doi.org/10.1016/j.stueduc.2013.10.001

Gibran, K. (1926). *Sand and foam*. Alfred A. Knopf. (Work includes the aphorism "Fear is a greater enemy than danger.")

Greenleaf, R. K. (2002). *Servant leadership: A journey into the nature of legitimate power and greatness*. Paulist Press. (Original work published 1977)

Hammoud, R., Zhao, Y., Selman, L. E., Chaskel, R., Triguero-Mas, M., Bilal, U., & Nieuwenhuijsen, M. (2022). *Birds' singing, people's wellbeing and life satisfaction: Real-time investigation using smartphone ecological momentary assessment. Scientific Reports, 12*, 11948.

Harari, Y. N. (2015). *Sapiens: A brief history of humankind*. Harper.

Hartig, T., Mitchell, R., de Vries, S., & Frumkin, H. (2014). Nature and health. *Annual Review of Public Health, 35*, 207–228. https://doi.org/10.1146/annurev-publhealth-032013-182443

Hoffman, E. (2007). *Foundations of existential psychotherapy: The humanistic-existential perspective*. Routledge.

Kaplan, S. (1995). The restorative benefits of nature: Toward an integrative framework. *Journal of Environmental Psychology, 15*(3), 169–182.

Kaplan, R., & Kaplan, S. (1989). *The experience of nature: A psychological perspective*. Cambridge University Press.

References

Keltner, D. (2023). *Awe: The new science of everyday wonder and how it can transform your life*. Penguin Press.

Kimmerer, R. W. (2013). *Braiding sweetgrass: Indigenous wisdom, scientific knowledge, and the teachings of plants*. Milkweed Editions.

Li, Q. (2018). *Forest bathing: How trees can help you find health and happiness*. Viking.

Louv, R. (2005). *Last child in the woods: Saving our children from nature-deficit disorder*. Algonquin Books.

Marris, E. (2011). *Rambunctious garden: Saving nature in a post-wild world*. Bloomsbury.

Metzner, R. (1999). *Green psychology: Transforming our relationship to the Earth*. Park Street Press.

Mitchell, R., & Popham, F. (2008). Effect of exposure to natural environment on health inequalities: An observational population study. *The Lancet, 372*(9650), 1655–1660.

Mortali, M. (2022). *Rewilding: Meditations, practices, and skills for awakening in nature*. Sounds True.

Murphy, L. (2024). *Reclaiming healing spaces: A phenomenological study on the transformative power of outdoor therapy from the lived experiences of Black clinicians working with Black clients* (Clinical doctoral project/dissertation, National Louis University). National Louis University Digital Commons.

Norton, T. A. (2016). Nature relatedness and experiential avoidance: A relevant connection? *Journal of Humanistic Psychology, 56*(5), 541–566.

Passmore, H.-A., & Holder, M. D. (2017). Noticing nature: Individual and social benefits of a two-week intervention. *The Journal of Positive Psychology, 12*(6), 537–546. https://doi.org/10.1080/17439760.2016.1221126

Passmore, H.-A., Garland, R., & Holder, M. D. (2022). Wellbeing in winter: Testing the noticing nature intervention. *Frontiers in Psychology, 13*, 840273. https://doi.org/10.3389/fpsyg.2022.840273

Seligman, M. E. P. (2011). *Flourish: A visionary new understanding of happiness and well-being.* Free Press.

Sullivan, W. C., & Kuo, F. E. (2011). *Do trees strengthen urban communities, reduce crime, and improve health?* U.S. Forest Service.

Tennessen, C. M., & Cimprich, B. (1995). Views to nature: Effects on attention. *Journal of Environmental Psychology, 15*(1), 77–85. https://doi.org/10.1016/0272-4944(95)90016-0

U.S. Forest Service. (2018). *Urban nature for human health and well-being: A research summary for communicating the health benefits of urban trees and green space.* U.S. Department of Agriculture. https://www.fs.usda.gov/sites/default/files/fs_media/fs_document/urbannatureforhumanhealthandwellbeing_508_01_30_18.pdf

References

Van den Berg, A. E., Maas, J., Verheij, R. A., & Groenewegen, P. P. (2010). Green space as a buffer between stressful life events and health. *Social Science & Medicine, 70*(8), 1203–1210. https://doi.org/10.1016/j.socscimed.2010.01.002

Wampold, B. E. (2015). *The great psychotherapy debate: The evidence for what makes psychotherapy work* (2nd ed.). Routledge.

Weller, F. (2015). *The wild edge of sorrow: Rituals of renewal and the sacred work of grief.* North Atlantic Books.

Wessels, T. (2005). *Reading the forested landscape: A natural history of New England.* Countryman Press.

Williams, F. (2022). *Heartbreak: A Personal and Scientific Journey.* W.W. Norton & Company.

Wilson, E. O. (1984). *Biophilia.* Harvard University Press.

Wolf, K. L., Krueger, S., & Flora, K. (2014). Healing and therapy—A literature review. In *Green Cities: Good Health.* College of the Environment, University of Washington. https://depts.washington.edu/hhwb/Thm_Healing.html

www.ingramcontent.com/pod-product-compliance
Lightning Source LLC
Chambersburg PA
CBHW051503030726

47592CB00006B/2067